WELCOME

The modern world depends on satellites for many services, but a transformation has torn out the roots of how they are designed, built, launched, and managed. Wealthy entrepreneurs have shaken the old order, creating a New Space age of commercial activity to accelerate the pace of change through services we all depend on. Gone are the old monopolies from established operators as disruptors such as Sir Richard Branson, Elon Musk, Jeff Bezos, and others are replacing traditional government funding with a self-sustaining programme, providing access to space far cheaper than it has ever been before.

From space tourism to routinely hauling satellites and spacecraft into orbit, the New Space era of privately run companies has up-ended the business practices of manufacturers and launch providers around the world. With launch costs going down, access to space is now easier for a wider range of companies providing services that power society today. And the revolution is extending beyond the practical benefits from communications, TV, navigation, and environmental monitoring. The goal of landing humans back on the Moon and using its natural resources to ensure a self-sustaining research base is now closer than ever through the international partnerships which underpin the commercialisation of space.

In this publication we learn about the existing order based on government-funded projects, track the start-up activities for space tourism, and follow the progress of entrepreneurs who have taken over the launch market for sending payloads into space. We also delve into the reasons why these innovators, using their own wealth, have been able to fast-track solutions to seemingly insoluble challenges and build rockets which can bring themselves back down for multiple re-uses. There are lessons here for the big aerospace corporations, losing out to creative new ways of managing large projects, start-ups succeeding where others have failed.

With opportunities for increasing numbers of people to travel in space, what are the

physical and psychological challenges? How easy is it to get a ride and what does it cost? And if you want to become an astronaut, how much do you get paid to do that job, an out-of-this-world opportunity to see the planet as relatively few had the chance to do previously? Answers in the following pages, as we track the milestones along this revolutionary road to the future where space travel becomes commonplace.

The story is a fascinating insight into how the established order of space-faring nations is shifting the balance from satellites and spacecraft being icons of a power-base to factors in a new, cheaper, and more beneficial use of national space programmes to embrace cooperation around the world. Enabled by commercial operators, humans are going back to the Moon as part of an endeavour supported by 40 countries, a capability which was unaffordable when tax-payers financed the effort to expand a human presence beyond Earth.

Welcome to Commercial Space, the new frontier of human endeavour unfettered from funding constraints binding public coffers, transformed into a self-financing enterprise bringing benefits and new opportunities.

David Baker
Author

ABOVE • Virgin Galactic gives thrill-seekers a view literally out of this world. (Virgin Galactic)

COVER IMAGE • Helmet illustration by Alexey Kotelnikov / Alamy Stock Photo/ Using elements supplied by NASA.

BELOW • A SpaceX Starship, the key to humans getting back on the Moon. (SpaceX)

44

58

CONTENTS

90

96

76

ISBN: 978 1 80282 983 9
Editor: David Baker
Senior editor, specials: Roger Mortimer
Email: roger.mortimer@keypublishing.com
Cover Design: Steve Donovan
Design: SJmagic DESIGN SERVICES, India
Advertising Sales Manager: Sam Clark
Email: sam.clark@keypublishing.com
Tel: 01780 755131
Advertising Production: Becky Antoniades
Email: Rebecca.antoniades@keypublishing.com

SUBSCRIPTION/MAIL ORDER
Key Publishing Ltd, PO Box 300, Stamford,
Lincs, PE9 1NA
Tel: 01780 480404
Subscriptions email: subs@keypublishing.com
Mail Order email: orders@keypublishing.com
Website: www.keypublishing.com/shop

PUBLISHING
Group CEO and Publisher: Adrian Cox

Published by
Key Publishing Ltd, PO Box 100, Stamford,
Lincs, PE9 1XQ
Tel: 01780 755131
Website: www.keypublishing.com

PRINTING
Precision Colour Printing Ltd, Haldane,
Halesfield 1, Telford, Shropshire.
TF7 4QQ

DISTRIBUTION
Seymour Distribution Ltd, 2 Poultry Avenue,
London, EC1A 9PU

Enquiries Line: 02074 294000.

KEY Publishing

BIG SPACE

From the Earth to the Moon

Long before the Space Age began, Cold War tensions between the United States and the Soviet Union pushed development of long-range ballistic missiles which could be used to strike each other with nuclear warheads. A lot of this technology grew out of German experiments during World War Two (1939-1945) and the development of the V2 rocket by scientists in Nazi Germany. Used against London, England, and Antwerp, Belgium, from September 1944, they were in reality the first objects to travel through space as they flew arching trajectories to their targets.

It had taken 10 years to develop these 'vengeance' weapons, which aimed to promote fear and panic in urban populations. But these rockets were not the only weapons to carry the 'V' symbol, the V1 being a flying-bomb powered by a pulse-jet engine. Unlike the V2 which could reach speeds of more than 3,000mph (4,800kph), the V1 flew within the atmosphere at a top speed of up to 400mph (640kph). The technology for these weapons was seized by the Allies and examples were delivered to research facilities in America and Russia, playing a significant role in the development of ballistic rockets and cruise missiles. But it was with rockets that the Space Age grew from fictional speculation to reality.

Ten years after the end of World War Two, rocket development reached a point where intercontinental ballistic missiles (ICBMs) could be developed with the capacity for America and Russia to attack each other with nuclear warheads. With flight times of 30 minutes or less, these weapons would have great tactical advantage over the use of strategic bombers taking several hours to reach their targets across hostile and heavily defended territory. By the late 1950s, Russia and America began testing their first-generation ICBMs.

The ideological and political differences between the two countries led to a competition to outdo each other in science, technology, and in demonstrating a superiority in the way their respective societies functioned and how the leadership could support the needs and aspirations of its citizens. World War Two had begun the process of liberation for many colonies of former empires and loyalty from these countries was sought by each side in the struggle for supremacy on the world stage.

But competition was not only in the field of military power and strategic capabilities for waging conflict. With a new range of post-war international organisations supporting non-military activities, specialist representatives from the two opposing regimes were frequently brought together at meetings and in conventions to agree peaceful and scientific matters, each sparring against the other for prestige and propaganda. One such was the re-establishment of the International Geophysical Year (IGY), which was to be held between 1957 and 1958 in the scientific survey of the Earth and its environment, providing a study of how the planet was influenced by activities on the Sun and its interaction with the atmosphere.

The IGY was to involve many countries around the world, an activity endorsed by the newly established United Nations, which placed emphasis on an equitable sharing of knowledge, information, and research opportunities for scientists from many nations. The Americans announced that they were going to launch a satellite to carry instruments and measure the region around Earth where the atmosphere interacts with solar particles and how it is affected by cosmic rays streaming through the solar system.

Rockets had been used for several years, carrying packages of instruments which conducted observations high above the atmosphere for a few minutes before falling back

BELOW • In the United States, the army took V2 rockets from Germany and launched them on experimental flights for research, this being the first launched from Cape Canaveral on July 24, 1950. (US Army)

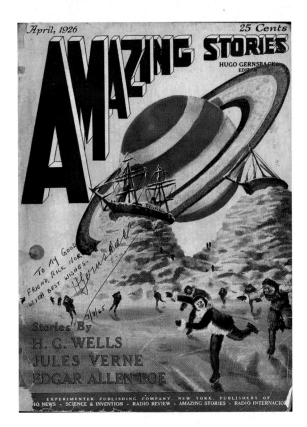

to the ground; but a satellite would remain orbiting the Earth for many years and could provide important information for scientists around the world. Such a capability would indeed be a demonstrable display of technological capability as well as placing the host country in a position of power and influence, more so perhaps than military prowess could. Recognising the value such a programme would have on their own image, as perceived by other countries, Russia announced that it too would launch a satellite into space.

The Americans had further reason for supporting its IGY satellite called Vanguard, in that it already had a spy satellite programme which was expected to begin operating by taking pictures of Soviet military bases and missile sites by the end of the 1950s. Scheduled to begin operations after Vanguard, the spy satellite had the code name Corona, but would be publicly declared as a research programme called Discoverer. At the time of its development, it was a highly classified project but a pioneer in its own right.

Under President Dwight D Eisenhower, the American government had been concerned about potential aggression from Russia after sending a satellite over Russian airspace – the Soviets had refused an American 'open skies' plan in which aircraft could routinely fly over each other's territory. The Vanguard, a scientific precursor to Corona, would help smooth the way for routine satellite activity, whatever its purpose. And in launching their own IGY satellite, the Russians too would be doing the same to America, although it would be some time before they had their own spy satellite.

Setting the Stage

The Russians had a different reason for wanting to display to the world that they were a technologically advanced society which could compete with America at any level. Having developed their long-range ICBM since 1953, Soviet Premier Nikita Khrushchev was persuaded by Russian scientist Sergei Korolev that by sending a satellite into orbit on a specially adapted version of this rocket, they would be demonstrating not only their military capabilities, but their technical prowess too.

The Soviet rocket would be known in the West as the R-7 Semyorka, eventually designated by NATO as the SS-6 *Sapwood*. The American Vanguard programme was proceeding at a somewhat leisurely pace while Korolev was given resources to place a Russian satellite in orbit as quickly as possible. The R-7 rocket specific to the satellite launch was far larger than the Vanguard rocket, which could place in orbit satellites weighing up to 2,900lb (1,315kg) compared with 25lb (11.3kg) for the American rocket.

Sputnik 1, the world's first satellite, was placed in orbit on October 4, 1957. It weighed 184lb (83.6kg) while the second and third satellites launched by Russia would realise the full lifting potential of the R-7. The American attempt on December 6, 1957 failed when Vanguard blew up on the launch pad. It was an embarrassing demonstration of how the Russians were already ahead of the United States in the public and open display of a technological race which henceforth would fund a space race that neither side had fully expected. A modicum of national pride was restored when the US Army stepped in and launched Explorer 1 on January 31, 1958, America's first satellite.

Neither country had anticipated the fear Sputnik would induce among the American public, well aware now that the Russians could attack the United States with impunity using rockets that were already demonstrating a superior capability through their space programme. But this was not quite the full story. Unaware of the Corona spy-satellite programme and of the emerging success the US Air Force would have with its own ICBM – the Atlas missile – the American public

LEFT • Imagination anticipated reality with early portrayals of the adventures humans might find in space, stimulating enduring ambitions to reach for the stars. (Author's archive)

BELOW • The great leap in rocket design came with the German V2 ballistic missile of 1944-45 which had a range of approximately 200 miles (320km) carrying a one tonne warhead. (Eberhard Marx)

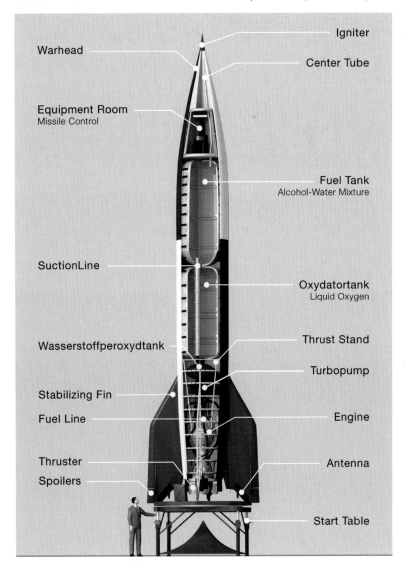

Warhead

Equipment Room
Missile Control

SuctionLine

Wasserstoffperoxydtank

Stabilizing Fin

Fuel Line

Thruster

Spoilers

Igniter

Center Tube

Fuel Tank
Alcohol-Water Mixture

Oxydatortank
Liquid Oxygen

Thrust Stand

Turbopump

Engine

Antenna

Start Table

ABOVE • *The Russians also got V2 rockets and trucked them back to Kapustin Yar in the Soviet Union along with German scientists and engineers to help develop their own programmes. (USSR Ministry of Defence)*

RIGHT • *The R-7 rocket developed as Russia's first ICBM and the launch system for its early satellites including Sputnik 1 launched on October 4, 1957. (Heriberto Arribas Arbato)*

BELOW • *Russian rockets developed for scientific research underpinned a range of missiles developed for the Soviet Army, here seen are the R-2A (left) with a dog capsule and the R-5. (USSR Ministry of Defence)*

demanded an appropriate response. Nothing less than the formal declaration of a government-led space programme to seize back the initiative and to achieve, and hold, leadership in satellites and space exploration.

Following one of the most intensively debated issues short of all-out war, the US Congress held numerous public and closed hearings after which it was announced that leadership in space and the public dissemination of all information it acquired, would be in the hands of a new civilian agency, the National Aeronautics and Space Administration (NASA). All military space programmes, of which there were now an increasing number, would be the responsibility of the US Air Force and a new government body, the Advanced Research Projects Agency (ARPA) which came into existence on February 7, 1958 followed by NASA on October 1.

The most visible change of pace was displayed by NASA when it formally announced plans for a range of Earth-orbiting satellites, spacecraft to fly by Venus and Mars, and even plans to put a man in space. Having secured the services of German V2 rocket engineer Wernher von Braun, the US Army had developed plans for a super-rocket to place very large objects in orbit and the US Air Force planned winged spaceplanes supporting a military role. The von Braun rocket would eventually be named Saturn and the initial development project for studying human physiology in space was handed to NASA and would be named the Mercury spacecraft.

It was in human spaceflight that NASA received the most interest from an American public hungry for space 'firsts' and a restoration of national pride. Congress was in charge

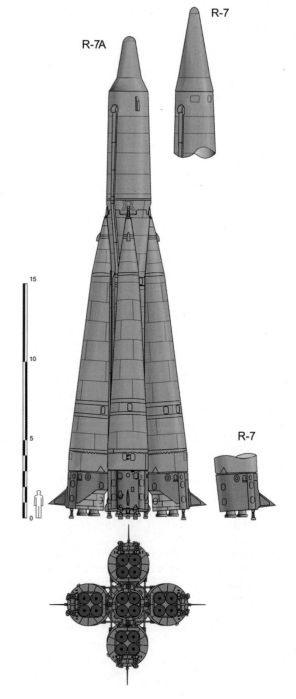

R-7

R-7A

R-7

of the public purse strings, for both the civilian and military projects, and voted a significant increase in NASA funding and for air force space programmes. The Cold War was being redefined as a struggle for ideological supremacy, non-committed countries watching as each side fought to outdo each other in the race for space.

In several respects the Americans had the clear and distinct advantage in human and financial resources, a now highly centralised and focused strategy for achieving more 'firsts' and in mobilising an industrial base which had so recently been hailed as the wartime 'arsenal of democracy'. The situation in Russia was very different, the full spectrum of space activities there during the early 1960s being controlled by the military, any non-military flights and project development having to satisfy the Moscow-based Ministry of Defence that it had either profound propaganda value or could in some way add value to military space projects.

A classic example of that was the development of the manned Vostok spacecraft which was only approved and funded on the basis that it also had the capacity to carry cameras and to operate as a spy satellite, for which it was known as Zenit. It had only been because of the persuasive powers of Sergei Korolev that Khrushchev had approved Sputnik at all, but the Soviet military was a persistent restraint on what the Russian space programme could have achieved had a long-term plan been supported with adequate funding and the necessary facilities.

From Earth to the Moon

Despite difficulties and the fight for resources, Korolev presided over the launch of Yuri Gagarin on April 12, 1961 – the first human in space – who was sent up on a developed version of the R-7 with an added upper stage and who returned to Earth on a parachute, having ejected from his Vostok 1 spacecraft on the way back through the lower part of the atmosphere. This alone added further incentive for US government support to the manned Mercury programme which saw Alan B Shepard become the first American in space on May 5, 1961, albeit on a short ballistic fight lasting little more than 15 minutes. It would be a further nine months before John Glenn successfully orbited the Earth three times on February 20, 1962.

John F Kennedy had been inaugurated as the 35th President of the United States on January 20, 1961, and the flight of Yuri Gagarin brought pressure on him to respond with a bold initiative to eclipse Soviet achievements in space. His ambitions were announced publicly before a joint session of Congress on May 25 when he tasked NASA with "landing a man on the Moon and returning him safely to the Earth." He set a high bar as an inspiration to America's youth to "do great things" and to engage with new challenges and, as he would claim in a speech in 1962, "we choose to go to the Moon in this decade and do the other things, not because they are easy but because they are hard; because that goal will serve to organise and measure the best of our energies."

In the pursuit of this ambition, the NASA budget grew dramatically over the next several years as Earth-orbiting research and science satellites were launched, spacecraft were sent to the nearest planets, and robotic precursors to the manned Moon missions served as pathfinders to the lunar surface. Four manned Mercury missions were flown before

ABOVE • America's Vanguard rocket and satellite programme developed for the International Geophysical Year (IGY) of 1957-58. (NACA)

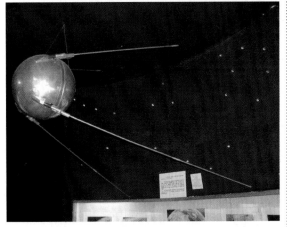

LEFT • A replica of Sputnik 1 at the St Petersburg Museum of Space and Missile Technology. (Andrew Butko)

the 10 manned Gemini flights of 1965 and 1966 demonstrated spacewalks, rendezvous and docking of two spacecraft in orbit, and extended flights of up to two weeks in duration.

The sequential steps necessary to get two men on the Moon by the end of the decade supported unprecedented growth for a peacetime enterprise, the workforce on space projects growing from around 2,000 when Russia launched Sputnik 1 in 1957 to around 400,000 by 1966. The number of satellites successfully launched each year by the US grew from five in 1958 to almost 100 by 1966, while the Russian total grew from one to 51 in the same period. The Americans had bounced back in force, between 1958 and 1966 successfully launching 437 satellites and spacecraft to Russia's 217, while the amount of money the US government spent on space projects grew from an authorised $347m in 1958 to $7,008m in 1966.

By this date America had overtaken Russia in space 'firsts' and overall achievements including flying the first three-man crew in 1964 and achieving the first spacewalk in March 1965. Not until 1969 would Russia successfully launch another cosmonaut into orbit. Before that, in January 1967 America lost three astronauts in a spacecraft fire at Cape Canaveral several weeks before their planned flight, and the Russians lost Vladimir Komarov, killed when his Soyuz spacecraft crashed to Earth that April.

By the end of the following year, NASA had test flown its new Apollo spacecraft, improved after the tragedy of the

fire, and sent three astronauts into an orbit about the Moon and returned them to Earth at Christmastime. The massive Saturn V rocket developed specifically to launch Apollo Moon missions had its precursor in the Saturn I series, first launched in October 1961 and which, by 1964, had equalled Russia's payload-carrying capacity. In achieving that, in the public mind the Americans had won the space race and NASA had put the United States ahead of the Soviet Union. But space was getting expensive and already the signs were clear that future plans and expectations would be moderated by costs.

NASA's Apollo programme had been put together by the government space agency but largely contracted out to aerospace companies, manufacturers familiar with building high-performance aircraft and powerful rockets. Primarily among which were Boeing, Martin, North American Aviation, Grumman, and Lockheed. Rocket engine manufacturers such as Pratt & Whitney and Rocketdyne built the motors which powered launch vehicles and spacecraft off the Earth and into space and the taxpayer voted in those who signed the cheques. No money went into space but rather into the pay packets of a skilled and semi-skilled workforce whose knowledge and expertise built the equipment behind the headlines.

It was, many believed, money well spent. Regarded as being on a 'war-footing', the pace and sheer energy behind the space programmes managed by Russia and America were unlike anything seen before in peacetime and the national pride they engendered in their populations grew with every passing 'first'. As the world watched, these two giant, nuclear-armed superpowers shaped technology and forged achievements on a superhuman scale, recruiting support as they sent their spacemen, and a Russian space woman, around the world on diplomatic handshakes.

In the Soviet Union, postage stamps carried commemorations of great deeds beyond Earth, posters everywhere proclaimed the winning ways of Marxist-Leninism and books extolled the benefits of the Space

Age in narrative form while films and documentaries brought dramatic coverage. To the outside world, Radio Moscow sang the praises of Soviet cosmonauts and news agencies such as TASS and Novosti handed out pictures and press releases. Throughout the communist world, space and the achievements of its cosmonauts rallied support and proclaimed Russia as a global player on the world stage.

The Morning After

But it was the landing of Neil Armstrong and Buzz Aldrin on the Moon on July 20, 1969 that threatened the future existence of NASA and the US space programme as Americans knew it. With a war raging in far-away Vietnam and economic pressures at home, the sheer cost of the space programme had been an accepted reality in achieving a propaganda coup against a country ruled by communist ideology, but Americans began to question the true value after the goal had been achieved.

Many had questioned it from the beginning, with 58% opposed to the idea of Moon landings the day after Kennedy had announced it in 1961, but they had grown to regard it as a national necessity and had watched with pride as the stars and stripes were planted in the lunar dust. Now there was uncertainty as to why they were doing it at all, politicians reminding space scientists, engineers, and NASA advocates, that they had funded a race and not a marathon with an undefined end point.

The cost of space exploration was largely in the price of rockets required to get off the Earth and that had been exemplified by the largest rocket of them all, the giant Saturn V which stood 363ft (110m) tall and weighed more than 3,000 tonnes at lift-off. Impressive in performance, each took more than two years to build and cost around $190m, around $1.9bn today. In fact, the rockets to get astronauts to the Moon cost more than the Apollo spacecraft to reach lunar orbit and the Lunar Module to go down to the surface and back combined. But NASA wanted more and lobbied for funds to keep Apollo going beyond the missions it could fly with leftover hardware.

That money never came, and Moon missions ended with Apollo 17 in December 1972. As the NASA budget declined, President Richard M Nixon contemplated abolishing NASA and putting its resources into a new government research and development agency to tackle national and global problems but that never happened. Instead, in 1973 NASA launched the Skylab space station assembled from existing hardware and sent three astronaut crews to conduct extensive science experiments in record-breaking flights in which the final crew spent 84 days in space during 1974. The last Apollo mission of all carried three NASA astronauts to meet and greet two Soviet cosmonauts launched in a Soyuz spacecraft during June 1975 as a goodwill mission to seal a new era of detente.

NASA had experienced funding cutbacks since 1966, when most development of hardware and the construction of all the necessary facilities had been completed and paid for. While most of the resources and the manpower had

ABOVE • The Apollo 11 spacecraft (right) is prepared for flight with Apollo 12 in the foreground, a programme that propelled the United States ahead of the Soviet Union in the Space Race. (NASA)

LEFT • Astronauts train for Moon landings and learn procedures for setting out scientific instruments on the lunar surface. (NASA)

RIGHT • Neil Armstrong (right) and Buzz Aldrin evaluate procedures for conducting geological tasks on the Moon. (NASA)

BELOW • Three of the six Moon landings had a Lunar Roving Vehicle for moving astronauts and equipment around between survey sites on the lunar surface. (NASA)

gone on the general expansion of the space programme - on its manned flights and into planetary exploration, scientific satellites, and orbiting observatories - small groups stood aside from the day-to-day running of existing programmes to look far into the future and provide options from which the NASA leadership could make recommendations to the White House for policy planning.

As a government agency itself, NASA cannot do anything without the approval of the White House, and the cabinet and its planning is based on what it is allowed to work toward, the goals and the programmes it would like to pursue and the amount of money it is allocated. Those funds are balanced by the US treasury on what the government as a whole can ask Congress to approve. The sequence is often protracted and requires a politically-savvy leadership at NASA to manage its operations in the best interests of the agency.

Sensing that the public was concerned over the escalating cost of the space programme, by the mid-1960s US politicians began to question the money it took to achieve the Moon goal and as early as mid-1963, several months before his assassination, Kennedy had himself pondered whether it was necessary to keep the race going all the way to the finish line. In just those two years since he came to power in 1961, great things had already been achieved by NASA and the Russians seemed, to the public and to many US politicians that they were not immediately heading for the Moon. As is now known, not until 1964 did they authorise an attempt to get to the lunar surface before the Americans.

In the aftermath of Kennedy's assassination, the American population rallied around the goals and the aspirations of the late President. His successor, Lyndon B Johnson, who had been the architect of NASA's formation in 1958 and of manoeuvring the Moon option into a declared goal, fully endorsed that objective. The Johnson administration had supported the push to beat the Russians and defended its priority in Congressional budget hearings but with high costs of the Vietnam war and civil unrest at home, the energy that had once powered Johnson's drive for a vibrant space initiative fell prey to other priorities, not least his domestic welfare plan.

Not until the Moon landing in 1969 did those levels of approval begin to change again quite radically. After Richard Nixon became President in January 1969, Vice President Spiro Agnew supported reinvigoration of national space goals by encouraging a root-and-branch examination of options for the future. Johnson had been unable to secure for NASA the funding necessary to keep production lines open for sustained Apollo missions and the costs in continuing with expensive, expendable launch vehicles such as Saturn ended all hope of the agency building more in a period of declining space budgets.

The End of a Beginning

When the results from NASA's in-house study were publicised in September 1969, what the agency called its Integrated Space Program was clearly a radical transformation from anything that had gone before, a key to which was a reusable spaceplane which could be used many times for sending satellites, spacecraft, and cargo into orbit and for retrieving worn-out hardware for repair and relaunch. Nuclear-powered rocket stages stored in orbit would be used to move equipment between Earth and the Moon, where reusable landers would support modest research stations and facilities where scientists could live and work for extended periods.

But the key enabling factor was that the spaceplane carry a crew to work in space, building orbital stations, servicing satellites, and supporting what NASA called 'in-orbit infrastructure', sets of equipment which would remain in space and be used as necessary for many different tasks. In developing a reusable launch system which could replace all existing rockets then in the NASA inventory for launching satellites and spacecraft, costs would topple, spacecraft would become cheaper because they could be retrieved and repaired, or worked on in orbit by teams of astronauts, and much more could be done with a lot less money.

The underpinning rationale was reusability and diversity: Reusability in that spacecraft and satellites could be flown many times without replacement and diversity in that each element within the system could be used for different purposes, reducing the number of separate vehicles that were required. A 'space tug' would be carried inside the spaceplane's cargo bay and deployed

in orbit to fetch and carry satellites from many different orbits and bring them back to the spaceplane where they could be repaired and returned to their appropriate location by the same tug.

To service the requirements of the Integrated Space Program, NASA discussed with industrial contractors - companies that until this time had built production-line satellites and space vehicles which could be used only once - ways to produce a more economical means of delivering the same services and extending the range of mission options available. It envisaged production of a limited range of platforms, each called the 'bus', onto which could be attached specific instruments and devices according to the requirements of customers bringing their own 'payloads'.

In this way, separate production of 'bus' platforms and 'payload' containers could simplify the need to find the right kind of satellite for a particular customer, or 'user'. NASA saw itself as custodian of these separate but highly integrated strands of bus, payload, and user to make it all simpler, cheaper, faster, and more efficient. But the sheer scale of the Integrated Space Program meant that government funding for the agency would have to rise again to finance the development of the spaceplane, an orbiting station, the space tug itself and the nuclear upper stage.

When Agnew got hold of the report, he encouraged planning for manned missions to Mars using these separate elements and tried to sell it to President Nixon but that got nowhere, despite NASA's encouragement to lobby the White House and Congress to get that approved. In 1969, with several Moon missions still to complete, Mars missions were never a realistic possibility at that time but the rest of the programme looked achievable, and while NASA set about the last flights to the lunar surface, planning for Skylab and arranging for the joint docking with a Russian Soyuz spacecraft, it began the process of reducing costs by getting Congress to fund a new age of reusable and affordable space travel.

ABOVE • *The Apollo programme achieved the high watermark in US space accomplishments before a period of uncertainty as to future direction and what could be achieved with much less money. (NASA)*

LEFT • *Eugene Cernan, commander of Apollo 17, the last manned Moon mission, and the last man to step off the surface on December 14, 1972, before returning to Earth with fellow Moonwalker Harrison Schmitt and Apollo crewmember Ronald Evans. (NASA)*

TICKET TO RIDE

Big ambitions and bigger rockets

To the general public, the end of NASA's Moon missions in 1972 and the last flight in the Apollo programme three years later were high points in NASA's meteoric rise since it had been formed in 1958. In just 17 years it had taken the American dream into space, planted the flag of the United States in lunar dust at six separate and highly distinctive sites, launched the world's first revisited orbiting space station, and linked up with a Russian Soyuz spacecraft to seal an era of hope and détente in foreign policy. Additionally, it had sent spacecraft to Mercury, Venus, Mars, Jupiter, and Saturn and launched several hundred scientific satellites and observatories.

Behind the scenes, that sense of achievement was celebrated but defined within NASA as the end of an era and the beginning of a new Space Age, one in which reusability was the route to a cost-effective programme. It was to achieve that through the development of a reusable spaceplane, the Space Shuttle, the assembly in Earth orbit of a permanently manned space station, and the availability of a space tug to move satellites and spacecraft in and out of different orbits. Over time, NASA would build a rocket stage powered by a nuclear reactor through its NERVA (Nuclear Engine for Rocket Vehicle Application) programme which had been in development for several years.

The redefined space programme would postpone Moon bases, scientific research stations on the lunar surface, and ambitious missions sending astronauts to Mars until those goals could be properly costed and funded. Instead, NASA would focus on near-Earth orbital activities, building a sustainable programme of reusable vehicles adapted to a wide range of tasks. It all served the perceived need for a wider and more diverse use of existing resources and a broader and more focused programme involving not only other government agencies but also the scientific, research,

and academic world where space and the weightless environment it provided could benefit society on Earth.

Prior to this transformation, space had been for government bodies to exploit in scientific, technological, military, and foreign policy applications, using it to advance existing capabilities or for propaganda and national prestige purposes. With the Cold War on the cusp of moving from confrontation to co-existence, the reasons for the Space Race had run their course and were no longer quite as applicable in a world changed by necessity from the immediate risk of nuclear war to a greater emphasis on international cooperation and the use of science and technology for more egalitarian purposes.

Nevertheless, the old order was still running its course. The Russians had embarked on their own Moon programme with giant rockets and advanced space vehicles approved and funded. But inadequate overall resources, too much reliance on competitive design bureaus fighting it out between themselves, and inadequate national testing facilities compromised the effort. They had started too late and were never given the tools to complete the job ahead of the Americans. Instead, the Russians had turned to orbiting space stations and were on the brink of achieving great success which, over time, would stand them in good stead when they too accepted a truly international connection with the world outside the former Soviet borders.

But the transition to a new way of planning a future for the space programme was tipping toward a greater connection between the high-end of science and engineering and the declared needs of ordinary people. Where NASA had used dramatic 'firsts' and expensive Moon missions to recruit support for America's technological virility symbols, it was now directing future plans toward the use of these engineering

BELOW • Russia developed Salyut space stations, with multiple modules supplied on separate flights. Here is a replica without solar cell arrays, showing Soyuz crew modules at left and a Progress cargo-tanker at right. (Donald Montgomery)

capabilities in broader applications. And in turn that defined the way the future space programme would share its priorities in both exploration and applications, for it was within the latter category that NASA began to move away from awe-inspiring programmes toward a benefits-based plan.

Launching on Demand

The American space programme had been built on the availability of ballistic missiles adapted as satellite launchers, beginning with the Redstone, Jupiter, Thor, Atlas, and Titan missiles modified as launch vehicles for a wide range of satellites and spacecraft. A small launcher, named Scout, provided low-cost rides for small payloads and these were available for customers on a reimbursable basis. This was a scheme whereby the US government signed a contract with a user who paid for the costs of launching a rocket carrying its payload into orbit. It was not intended to make a profit, simply to reimburse the government for the cost of building the launcher and sending it into space.

Reimbursable launches were becoming standard practice for payload operators and users from many friendly countries, serving to broaden access to space and to provide a useful way for the United States to engage with foreign countries and governments. Soon, that deal would extend to private operators from research institutions, universities and organisations providing a platform for private companies to put individual experiments on satellites which were then sent into space. The Redstone and Jupiter rockets were used for early flights carrying US Army and US Air Force satellites and for NASA, but they were soon replaced by larger and more capable launch vehicles.

The Thor ballistic missile, 60 of which had been deployed to Britain with nuclear warheads between 1959 and 1963, was redeveloped by its manufacturer - Douglas Aircraft - as a satellite launcher with a small upper stage and named Delta. The first successful flight was achieved on the second launch which carried a science satellite for NASA on August 12, 1960. With a Thor-Able upper stage it could place a payload of 260lb (120kg) in orbit but later variants with more powerful upper stages and strap-on boosters grew in size and potential.

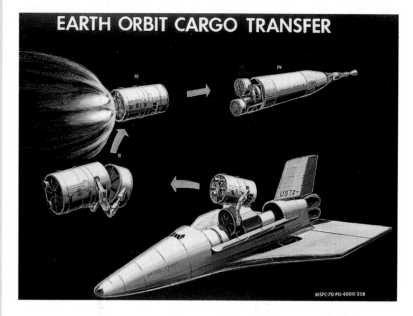

EARTH ORBIT CARGO TRANSFER

MSFC-70-PD-4000-25B

Not appropriate for reimbursable flights due to its size and cost, the much larger General Dynamics-built Atlas ICBM was readily available for production as a government satellite launcher and with its more powerful rocket motors it began to be used for NASA and US Air Force missions from 1959. The Atlas served only a brief time as part of the nuclear deterrent, a developed version serving as a satellite launcher. Equipped with a range of optional upper stages and a maximum payload capacity of around 4,100lb (1,860kg) to low Earth orbit, it could send a 2,300lb (1,040kg) spacecraft to the Moon.

Atlas with a Centaur upper stage formed the core of US space achievements in the 1960s and 1970s, launching a wide range of lunar and planetary missions as well as science satellites and solar observatories as well as the target vehicles for NASA's manned Gemini rendezvous and docking flights. Many would regard its seminal achievement to be sending six Surveyor soft-landing space probes to the surface of the Moon between 1966 and 1968, and for several Mars missions.

The Titan ICBM was built by The Martin Company and developed as a back-up to the technically challenging Atlas, using more conventional forms of rocket engineering and stage design. But it was slightly more powerful than Atlas

ABOVE • After Apollo, NASA looked to cut costs through reusability, using a shuttlecraft to support the use of a space tug and a nuclear upper stage to deliver large loads to the moon and Mars. (NASA)

BELOW • With the space tug, NASA hoped to operate a multipurpose vehicle capable of carrying crew, moving cargo between orbits, or taking supplies to a Moon base. (NASA)

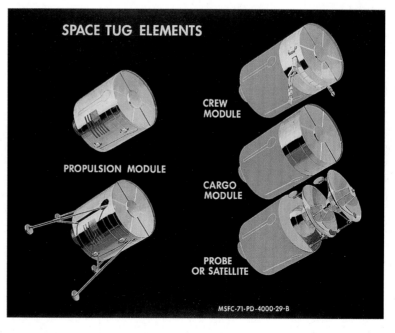

SPACE TUG ELEMENTS

CREW MODULE

PROPULSION MODULE

CARGO MODULE

PROBE OR SATELLITE

MSFC-71-PD-4000-29-B

and, when paired with appropriate upper stages, could carry heavier payloads into orbit. Its future lay in added solid-propellant boosters with two attached to the first stage and in this configuration as the Titan IIIA it began to launch satellites for the US Air Force, the first successful flight being on December 10, 1964. Various derivatives followed, with the powerful Titan IIIC supplemented by large strap-on boosters delivering a payload of 28,900lb (13,100kg) to low Earth orbit. That launcher made its first successful flight on June 18, 1965.

Most readers will be aware of the Saturn series of launch vehicles which underpinned the Apollo programme, and which successfully placed in orbit all the payloads assigned to them. Ten Saturn Is were launched between 1961 and 1965, the final six of which had an upper stage providing a payload capability of 20,000lb (9,072kg) to low Earth orbit. With a more powerful upper stage the Saturn IB had a lift capacity of 38,000lb (17,237kg) and, in addition to four unmanned flights beginning in 1966, it was used between 1968 and 1975 to launch the first manned Apollo mission, the three Skylab crew delivery flights and the joint docking mission with the Russia's Soyuz. The first stages for Saturn I and IB were built by Chrysler and the second stage by Douglas Aircraft powered by rocket motors from Rocketdyne.

The ultimate development was the three-stage Saturn V which made its first unmanned test flight in November 1967 followed by a second in April 1968. It carried ten manned Apollo missions between 1968 and 1972 and launched the Skylab space station in May 1973. It had a payload capacity of almost 312,000lb (141,523kg) to low Earth orbit and could

send 108,000lb (48,989kg) to the Moon. Several proposals were submitted for further development of the Saturn V, but they were expensive to launch and there were very few potential payloads to justify continuing the production line. Several manufacturers had been involved, Boeing building the first stage, North American Aviation building the second stage, and Douglas Aircraft the third stage with rocket motors again provided by Rocketdyne.

A Surge in Satellites

By the end of the Apollo programme the demand for launch vehicles was increasing at a faster rate than had been anticipated. Reimbursable launches were growing, as was the rush to commercialise space applications such as communications and TV relay services. Existing US companies wanted to launch satellites to broaden their markets and improve network coverage, to reduce the cost of expensive ground installations and to attract new markets on to their radio and TV stations as well as carrying telephone services. Telecommunication companies such as Bell and AT&T were at the forefront of a drive to give America pre-eminence in satellite relay across the Atlantic and around the world.

On September 25, 1961, President Kennedy announced that the government was setting up INTELSAT, the International Telecommunications Satellite organisation, in an agreement including seven foreign countries. INTELSAT was based on an equitable sharing of capacity based proportionately on the financial contribution each country made to running and operating the system. The more you paid the more you could have access to its bandwidth. It would launch its own satellites, the first of which was sent up on a Delta rocket in April 1965 known as Intelsat I, and popularly nicknamed Early Bird, it was positioned over the Atlantic Ocean for US-European coverage.

Other satellites followed and the system grew as the first widespread commercial application of the space programme. Coverage over the Indian and Pacific Oceans by mid-1969 carried the TV broadcast of the first Moonwalk live around the world. A system which was essentially owned by the countries which used its services, by 1973 there were 81 signatories to the INTELSAT agreement with each one being the designated telecommunications provider in that country. Most of the bigger satellites transitioned to the Atlas-Centaur rocket to take advantage of the heavier weight-lifting capacity of that launch vehicle.

The INTELSAT model would be copied in other sectors and the burgeoning demand for space applications began to outgrow the capacity of governments alone to provide the infrastructure and the means to deliver services

which private companies began to pick off and exploit. But the desire on the part of the US government to control the international telecommunications services brought restrictions in the launch of foreign satellites which threatened to compete with American companies and the interests of the commercial world.

As INTELSAT grew and expanded, control was still centralised in the United States, but other countries were developing their own space programmes which could be seen as competing with American interests. Europe had two separate and distinct organisations, one which sought to develop space for scientific research (ESRO) and another which aspired to build an independent launch capability (ELDO). ESRO – the European Space Research Organisation – was largely responsible for developing scientific satellites which were being launched by American rockets on a reimbursable basis, while ELDO – the European Launcher Development Organisation – would build European rockets for launching satellites.

The ELDO programme began when Britain encouraged Germany, France, and Italy to back a Europa launch vehicle programme, each country paying a proportional contribution and each developing a separate part of the rocket. Britain had a head start in rocket propulsion with development of a long-range ballistic missile, Blue Streak, which it began in 1955 in cooperation with America. When the UK government decided in 1960 that such a missile was too vulnerable to attack, that it was too expensive, and that the nation really didn't need such a deterrent, it abandoned Blue Streak as a weapon and went to Europe to propose it as the first stage for an independent launcher appropriately called Europa.

The ELDO agreement had France build the second stage, Germany the third stage and Italy the experimental satellite platform for initial test flights. France was developing an indigenous but small satellite launcher and would use that to place its own satellite in orbit in 1965 from a site in Algeria. Germany was already conducting experiments into propulsion systems for Europa, but the following year Britain made it plain that it wanted to leave the project and by 1970 it was clear that this also included withdrawing Blue Streak as the first stage and effectively torpedoing the Europa programme.

Development flights had begun in June 1964 with the launch of Blue Streak on its own from the Woomera Test Range in Australia, 10 flights having been completed by June 1970 of which five had upper stages. Except for the five solo Blue Streak launches, none of the multi-stage attempts were successful. France had been largely responsible for building the planned operational launch site at Kourou in French Guiana on the coast of South America and the only test flight from that location occurred in November 1971, but it too was a failure.

When Britain pulled Blue Streak out of the Europa launcher programme it did serious damage to its reputation on the continent, solidifying resolve in France and Germany for autonomy and, despite the UK reneging on its agreement, to continue with the programme. It mattered more because there were concerns that America would not allow reimbursable launches for satellites believed to compete with US interests, or organisations such as INTELSAT which it controlled as the majority stakeholder.

That issue came to a head with the request from Europe for the US to launch its Symphonie experimental telecommunications satellites built by a Franco-German consortium, but which were considered to be encroaching on American telecommunications services. Discussions

between the organisation and the US State Department resolved the issue when the Europeans declared that activities with Symphonie would not impinge on US services. But it solidified even further European resolve to operate an independent launch capability and that began the process of merging ESRO and ELDO into the European Space Agency (ESA) in 1975 and to proceed with development of the Ariane launch vehicle. Britain would play only a small part by providing the Ferranti guidance platform for the new vehicle but took no part in funding it.

As noted previously, along with all this growth in the use of space for commercial purposes and the potential for satellite launcher markets, NASA was being redirected by the US government to completely transform the American non-military space programme by abandoning projects based on expendable hardware for a sustainable age of affordable, reusable rockets and satellites. At the core would be the spaceplane and studies for that concept began in 1967 leading to what the space agency initially labelled to be an Integral Launch and Re-entry Vehicle (ILRV), essentially putting both spacecraft and launcher together in a combined system, all of which would be reusable and cost-effective. It would soon acquire a more marketable name – the Space Shuttle.

ABOVE • Developed as a ballistic missile, the Jupiter would also carry some satellites during its brief use in that role. (US Army)

Mars, and the adaptable space tug, NASA sought partners for initial studies. A range of aerospace companies familiar with building rockets and satellites began work on shuttle concepts while two major manufacturers, North American and McDonnell Douglas, studied space stations in parallel.

When these studies were conducted in the late 1960s, NASA still had hopes for continuing with Saturn V production and the space station was based on availability of that launcher, design proposals envisaging large cylindrical structures placed in orbit to which would be attached scientific modules also launched by expendable rockets.

The Shuttle, meanwhile, emerged as a truly integrated system consisting of a winged booster lifting a winged orbiter partway into space before returning to a piloted landing, the orbiter achieving orbit with its own rocket motors where it would carry crew and cargo to the space station before itself returning to land on a runway close to the launch site. There, it would be prepared for flights back into space using the same reusable booster.

NASA was not new to these concepts and had been formed in 1915 out of the existing National Advisory Committee for Aeronautics (NACA). Inspired by the British Advisory Committee for Aeronautics formed in 1909, the NACA was set up to conduct government-funded research into the scientific principles of flight and the practical applications of aerodynamic principles which it would research in its laboratories and wind tunnels. By the end of World War Two its activities were diverse and engaged with the basic principles of rocket propulsion and rocket-powered research aircraft.

Thus, it was that bringing satellite-launching rockets and winged spaceplanes together was merely a logical evolution from work it had already been conducting for some time. But the cost of a fully-reusable, fly-back system was too expensive for the diminished financial resources of NASA in the post-Apollo period and Congress refused to

ABOVE • The US Air Force developed the Thor missile deployed to the UK with a single nuclear warhead before it was adapted for satellite launches, the first of which took place in 1960. (USAF)

A Space Redux

Following the post-Apollo Integrated Space Program and its objectives for a reusable spaceplane, space station, nuclear rocket stage for fast flights between Earth and the Moon or

RIGHT • The configuration of a Thor rocket renamed Delta for its peaceful use as a satellite launcher, here in the role of sending a Telstar communication satellite into orbit in 1962. (NASA)

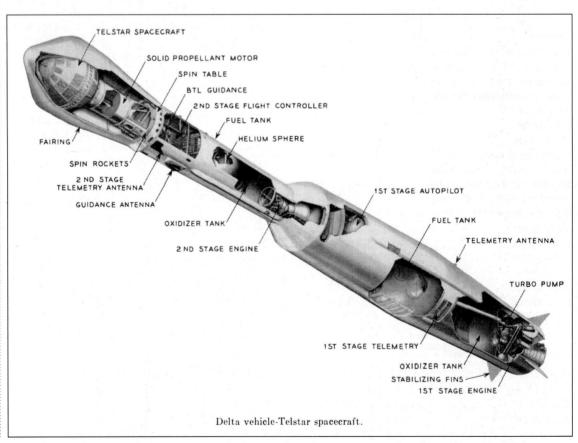

Delta vehicle-Telstar spacecraft.

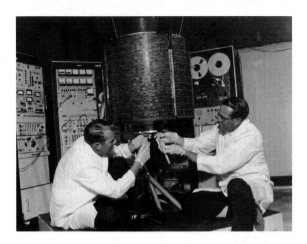

consider that fully reusable concept with fly-back booster and orbiter, forcing the agency through a lengthy series of further studies, analyses, and wind tunnel tests to find a less expensive solution.

After three years of additional study, it came up with a heavily compromised system involving a reusable winged orbiter launched on its way by two solid-propellant rocket boosters. Gone was the large, piloted, fly-back booster. The orbiter would be powered by three rocket motors in the tail fed with propellants stored in a large external tank to which the orbiter and the boosters would be attached. It was planned that the boosters would parachute back down into the ocean and be recovered where they could be used again after refurbishment, but the external propellant tank would be destroyed as it fell back down through the atmosphere.

In this form, the Shuttle was approved by President Nixon in January 1972 and funds released by Congress, justified on the basis that it would reduce the overall cost of the government space programme by achieving some level of reusability. Congress would have to apportion funds to build the Shuttle and to operate it, but over time the aim was to encourage a more sustainable approach by having common bus elements for payloads so that users could gain more cost-effective access to space.

By stimulating the private sector, NASA was continuing the original objective of its predecessor, the NACA, which existed to encourage and push industry to becoming a world leader in aviation and aeronautical engineering. The record showed that it had achieved that, by providing a catalogue of wing profiles and aerodynamic shapes for fuselages and appendages from which manufacturers could draw on as proven references, enabling design teams to fast-track the development of civilian and military aircraft. Many at NASA wanted to replicate that for the emerging space industry, to provide stimulation for private enterprise releasing the government from the burden of funding everything to do with space.

Throughout the 1970s, while the Shuttle was being developed and the hardware built, greater uses of space began to emerge to support the idea that this reusable space transportation system could encourage private enterprise by cutting the cost of sending satellites and experiments into space. Long before the Shuttle was approved by Congress, the parallel development of the Shuttle and the space station it had been conceived to service was deemed too costly and the agency shifted these two into separate, consecutive programmes, first the Shuttle and only then, when the peak of its development had been paid for, the station. But there was concern. Without guarantees that the station would ever be funded,

NASA began to broaden the way the Shuttle could be made to pay for itself.

Before the first Shuttle flight on April 12, 1981 – 20 years to the day after Yuri Gagarin made the first human space flight – NASA was committed to selling its services on the basis that it would be replacing all existing, expendable rockets. In would come the reusable spaceplane, out would go the Delta, Atlas, and Titan launch vehicles which had sent all American, and a few foreign, satellites into space. These expendable rockets had been revenue-earning projects for McDonnell Douglas, Convair, and The Martin Company respectively and now were to be condemned to history if customers previously sending their payloads on those launch vehicles could be persuaded to book aboard the Shuttle.

To satisfy the requirements of potential customers, NASA had to provide additional equipment capable of reaching required orbits. By the time flight operations began in the early1980s, the space tug had been abandoned as an unaffordable expense as had the

LEFT • Named Early Bird, INTELSAT 1 ushered in a new era of semi-commercial space operations when it was launched in 1965 at the start of a global telecommunications programme in which almost every country participated for relaying telephone and TV signals. (Hughes Aircraft Company)

LEFT • With Delta, growth was exponential in raising the payload capability through the addition of solid-propellant strap-on boosters to augment the first stage. (NASA)

BELOW • The high-energy, cryogenic Centaur upper stage was developed for Atlas and that produced a winning combination for launching a very wide range of satellites and spacecraft. (General Dynamics)

RIGHT • The Atlas, America's first intercontinental ballistic missile would have a brief operational deployment in its design role but achieve lasting fame when developed for the space programme. (USAF)

FAR RIGHT • The advanced INTELSAT IV-A of the mid-1970s significantly expanded coverage and capacity for the global telecommunications network, the first era of commercial space systems. (Hughes Aircraft Company)

BELOW • Britain abandoned its Blue Streak missile before deployment and offered it to Europe as the first stage of a European satellite launcher, then withdrew leaving Europe without an independent rocket to put satellites in orbit. (John McCullagh)

nuclear-powered upper stage. But an increasing number of commercial operators buying satellites for radio, television, and telephone services had satellites operate from geostationary orbits, a position 22,200 miles (35,880km) above the surface of the Earth and in the plane of the equator, where a full orbit takes exactly 24 hours, the same time the Earth takes to rotate once on its polar axis making the satellite appear stationary.

Geostationary orbits were the secret to continuous service from a single satellite, as it appeared to never move from its position in the sky, high above Earth. Although they were expendable, unmanned rockets could reach this height whereas the Shuttle could only reach altitudes of around 400 miles (643km). The system required a boost stage of some sort to move the satellites from an orbit the Shuttle could reach to one they were required to operate from. Three were available, all having been developed as upper stages for expendable rockets: The Inertial Upper Stage (IUS) and two different sizes of Payload Assist Module

(PAM), one for Delta rockets (PAM-D) and a slightly more powerful one for Atlas (PAM-A). All three could be adapted for the Shuttle.

Replacing Rockets

With no space station for it to service in sight for at least a decade after it made its first flight, emphasis switched to making the Shuttle the space truck to carry everything launched by the United States. That included retiring the Delta, Atlas, and Titan rockets which had been doing the heavy-haul since the dawn of the Space Age. The rocket manufacturers were making money, although not a lot of profit, from government orders for these expendable launch vehicles but during the 1970s, while the Shuttle was being built, countries around the world began increasingly to invest in satellites, primarily for telecommunications.

By the early 1980s when the Shuttle started flying, there was a buying frenzy of satellite services and American manufacturers began to reap real benefits from orders placed by many countries around the world. Commercial satellite services fed right into domestic broadcasting and orders began to stack up for standardised satellites. Canada was one of the first buying its satellites from US manufacturers, principally Hughes Aircraft Company at first, before Ford Aerospace, General Electric, and a few more got in on the act.

The breakdown of a satellite into separate bus and payload sections produced an economy of scale. A manufacturer would design a standard bus providing electrical power, station-keeping propulsion, communications systems for monitoring its operation, and command and control electronics. To these would be attached the customer's payload, such as the communications relay equipment for processing, boosting and relaying telephone and TV signals supplying services to its domestic customers. International traffic went by INTELSAT.

NASA got in on the act with the Shuttle, seeking to secure a monopoly on launch services by replacing all expendable rockets with the reusable system. By the early

P-51 VERSUS LUFTWAFFE

The German Luftwaffe, the air force that defended the Third Reich against the Allied air forces in World War Two, was an elite service arm that fielded state-of-the-art fighter aircraft flown by experienced combat airmen. Designed for the fighter-interceptor role, the primary aircraft of the Luftwaffe were the single-seat Messerschmitt Me-109 and the Focke Wulf Fw-190, both produced in large numbers and in numerous variants through the course of the war.

During the war's early years, Luftwaffe fighters performed air superiority sweeps and bomber escort duty in the West as the Battles of France and the epic air Battle of Britain were fought, in the East after the launch of Operation Barbarossa, the Nazi invasion of the Soviet Union in June 1941, and in the Mediterranean as the Desert War and Italian campaigns were prosecuted. However, as the tide of war turned against the Axis, Luftwaffe fighters assumed a more defensive role. They attacked waves of Royal Air Force and US Army Air Forces bombers that dropped tons of explosives on German military and industrial targets and population centres.

In time, the experience level of the German pilots dropped significantly due to combat attrition, while the number of available fighters was curtailed as well. At the same time, Allied fighter strength grew in numbers and in capability, as evidenced by the prowess of the North American P-51 Mustang. Still, German technology was impressive, and along with the Me-109 and the Fw-190, Luftwaffe pilots flew the Messerschmitt Me-262, the world's first operational jet fighter, the rocket-powered Me-163 Komet, and numerous twin-engine interceptors during the air war in Europe.

The Messerschmitt 109 was a workhorse of the Luftwaffe and one of the most famous aircraft types in the history of military aviation. It was originally designated the Bf-109 in

reference to its early manufacturer, the Bayerische Flugzeugwerke, and then after 1938 it was named the Me-109 in reference to its designer, the legendary Willy Messerschmitt, in co-operation with Walter Rethel and Robert Lusser, an employee of Bayerische Flugzeugwerke. Early design work was begun in 1934, and the first flight of the Bf-109 took place in September 1935. The type entered production in 1937 as the B-1, followed in 1939 by the Me-109E, 1940 the Me-109F, and 1942 the Me-109G as the primary variants.

During a production period that spanned 21 years, reaching well beyond the end of World War Two, the Me-109 was built by Messerschmitt AG in Germany, and

under licence by Dornier-Werke in Switzerland and Hispano Aviacion in Spain. More than 35,000 were manufactured, the last of these in 1956. In World War Two, more than 70 per cent of the Me-109s »

LEFT: Luftwaffe ground crewmen rest beside an Me-109E fighter of JG 53 (Jagdgeschwader or Fighter Wing 53) at an airfield in the West about 1939-1940. Creative Commons Bundesarchiv Bild via Wikipedia

LEFT: A flight of Me-109 fighters streaks across the sky. The iconic Messerschmitt was built in greater numbers than any other Luftwaffe fighter of World War Two. Creative Commons Bundesarchiv Bild via Wikipedia

BELOW: Luftwaffe ground crewmen push an Me-109G across an airfield in France in 1943. Creative Commons Bundesarchiv Bild via Wikipedia

received by the Luftwaffe were of the G, or Gustav, variant that was built from 1942 until the end of the conflict.

Powerful plane

The Me-109G was powered by a 1,475hp Daimler Benz 605A-1 V-12 inverted, liquid-cooled piston engine. Capable of a maximum speed of up to 428mph, it was one of the fastest aircraft of its time with a service ceiling of 38,000ft and range up to 460 miles. Its formidable armament included a mix of weapons in the G-series. These ranged from a pair of 13mm MG 131 machine guns in the engine cowling synchronised to fire through the propeller, a centreline 20mm MG 151/20 or 30mm MK 108 cannon firing centreline through the nose, two 20mm MG 151/20 cannon slung in underwing pods, 210mm Wfr Gr 21 rockets in wing mounts, and a single 551lb bomb or four 110lb bombs attached to wing hard points.

Relatively small at roughly three tons weight, wingspan of 32ft 6in, and length of 29ft 8in, the Me-109G was fast, easily mass produced, possessed an exceptional rate of climb and dive, and excellent manoeuvrability. Its deficiencies included narrow landing gear that was unforgiving to novice pilots, strong swing on take-off and landing – again difficult for

inexperienced pilots to handle – diminishing lateral control at high speed, and a tendency for wing slats to open during tight turns to prevent stalls but which threw the pilot's aim off at the same time.

Another primary weakness was relatively poor rear visibility from the cockpit, its rear flush with the high fuselage. The Me-109G was heavier than both the preceding E (Emil) and F models and exhibited characteristics that required the pilot's constant surveillance during flight, while landing the Gustav was a tremendous challenge often described by pilots as 'malicious'.

Most the Luftwaffe's successful aces of World War Two flew the

Me-109, and they testified to its high-performance capabilities. Among them were Major Erich 'Bubi' Hartmann, the highest scoring ace of all time with 352 aerial victories, Major Gerhard Barkhorn with 301 air kills, and Lieutenant General Adolf Galland, Luftwaffe General of Fighters, who flew 705 combat missions during his career from the Spanish Civil War to the end of World War Two and shot down 104 aircraft.

The Me-109 was the workhorse fighter of the Luftwaffe for the duration of World War Two, and Allied fighter pilots respected its capabilities in the hands of a skilled, veteran pilot.

The Focke Wulf Fw-190 'Shrike' was designed by the renowned company engineer and test pilot Kurt Tank. It first flew on June 1, 1939, and the Fw-190A-1 entered production in September 1940. More than 20,000 examples in numerous variants were built by the end of the war, and 65 were produced in France after the end of the conflict. The Fw-190 was configured in several fighter-interceptor and ground attack fighter-bomber versions, and its average empty weight did not exceed four tons. It was a nimble fighter, diminutive for its time with a wingspan of just under 35ft and length of 29ft. Powered by the BMW 801D-2 14-cylinder, air-cooled piston engine, it was capable of a top speed of 405mph, with a range of up to 620 miles, and a service ceiling of nearly 34,000ft.

Armament consisted of a variety of options. The Fw-190A-8 mounted a pair of 13mm machine guns in the engine cowling synchronised to fire through the propeller, two 20mm MG 151/20 cannon synchronised in the wing roots, two 20mm MG 151/20 cannon placed at mid-wing, and either a single bomb under the fuselage or four smaller explosives affixed to hard points on the wings.

ARIANE 1   ARIANE 2   ARIANE 3   ARIANE 4   ARIANE 5   ARIANE 5 Evolution

LEFT • In response to Britain pulling out of providing a European launch vehicle to make it independent of the Americans, France organised the Ariane programme and the formation of the European Space Agency. As displayed here, Ariane would grow and expand, launching a significant proportion of the world's satellite traffic. (Arianespace)

1980s, research organisations, scientific institutions or foreign countries were looking for a ride for their satellites, which were invariably bought from a US manufacturer. Without the big space station, now deferred until at least the 1990s, NASA played the role of service provider, selling payload space aboard the Shuttle.

Invariably, customers wanted to get their satellites into geostationary orbit which required them to be attached to an upper stage and carried in the Shuttle's spacious payload bay, which was 60ft (18.2m) in length and 15ft (4.4m) in diameter. With that volume and a total payload capability of about 50,000lb (22,680kg), the Shuttle could carry three or four separate satellites attached to their PAM-D boost stages. Once in orbit, the Shuttle's payload bay doors would open and each satellite and boost stage would be spring-ejected to drift away and, from a safe distance, fire the motor to push it all the way up to a geostationary position.

The first mission of this type came on the fifth flight of the Shuttle, launched on November 11, 1982 carrying communications satellites for Satellite Business Systems and Telesat Canada. The next flight on April 4, 1983 lifted into geostationary orbit a large US government relay satellite equipped to replace many ground stations by linking several separate satellites with a single point of reception on the ground from where the connections were routed to the separate users. This was followed on June 18 by the launch of another satellite for Canada and one for Indonesia, which would be used to connect the 6,000 occupied islands of that country to save the cost of wiring up each one with landlines. And so, it went on, with the next flight carrying a satellite for India bought from Ford Aerospace.

By 1984 the Shuttle was launching up to three satellites on some missions under a pricing structure offered to undercut that charged for flying on expendable rockets – market principles governing the attraction for customers induced to shift from expendables to the reusable system. In doing so, peripheral costs also began to come down. Insurance for satellites operating on a business structure for commercial companies takes account of reliability and the risk factor in probability of failure and the perception was that the Shuttle had a safer record and would be less likely to 'lose' a satellite. Confidence was dented, however, when two satellites failed and had to be retrieved by the Shuttle on a later mission, the insurers being an integral part of that operation.

BELOW • Early concepts for a reusable shuttle in 1970 anticipated fly-back booster and orbiter but the cost of that was itself too high. (North American Rockwell)

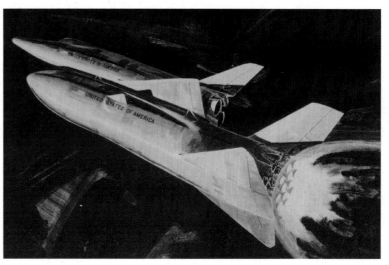

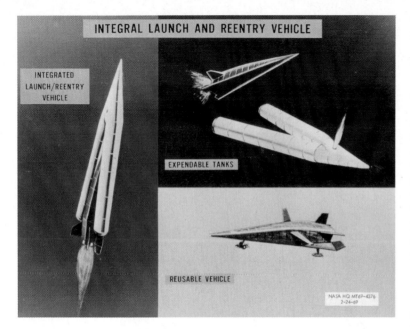

INTEGRAL LAUNCH AND REENTRY VEHICLE

INTEGRATED LAUNCH/REENTRY VEHICLE

EXPENDABLE TANKS

REUSABLE VEHICLE

NASA HQ MT69-4376
2-24-69

ABOVE • Seeking a more cost-effective way of running government-funded space projects, NASA studied an Integral Launch and Re-entry Vehicle concept, a post-Apollo shuttle which could replace expendable launch vehicles. (NASA)

NASA regulated the price charged for carrying non-government satellites into orbit in cooperation with the State Department, but this did not equal the cost of launching a Shuttle and, as the frequency with which the Shuttle could be launched was lower than expected, the price and the cost could not pay for the total accrued by the flight. When assessed back in 1971, the economic basis on which the Shuttle was approved assumed a potential for up to 60 flights a year for which the same programme with expendables was calculated to cost more. Anything less than 48 flights year and the Shuttle-based programme

would be more expensive. But the Shuttle was more complicated to operate and complex to maintain than had been believed when it was designed and never achieved more than nine flights a year.

With the exception of the Titan reserved by the US Air Force as back-up to any problems with the Shuttle programme, the entire expendable launch industry was in a state of shutdown when *Challenger* was destroyed shortly after lift-off on January 28, 1986. In the ensuing review of why that had happened, the Shuttle was delayed while modifications were made, and it would not return to flight before September 29, 1988. From that date it was retained for launching only government payloads and the near demise of the expendable launch industry went into immediate reverse and began to gear up to resume carrying the traffic it had before the Shuttle offered cut-price deals.

The *Challenger* review board concluded that the Shuttle programme had been under too much pressure and that rushing to satisfy commercial customers and accelerated demands to launch on time had tipped programme managers into risks that would not otherwise have been taken. Gone now was the revenue the Shuttle programme could receive from commercial satellite customers. From this point, the Shuttle would be used almost exclusively for assembling a space station for which Congress had granted funds for optional design studies.

Two years before *Challenger* transformed not only the spectrum of Shuttle mission payloads but the space programme as a whole, NASA had proposed a station called Freedom which would be assembled from separate modules launched by the Shuttle. The giant Saturn V had long gone and the only way to build a permanently

RIGHT • The shuttle that was approved for development in January 1972 had solid-propellant boosters and propellant for the winged orbiter in a disposable external tank to which those other elements were attached. (NASA)

manned research facility was to assemble it from separate sections like a giant Lego kit. But the budget for such a facility was more than NASA would be allowed from government coffers. The only option was to recruit international partners to share the cost by contributing separate research modules in return for which they would be allowed to fly on the Shuttle and conduct their own experiments at the station.

By the time of the *Challenger* disaster, Canada, Japan, and the European Space Agency had signed up to the station project but the collapse of the Soviet Union and the offer of participation in what was rebranded as the International Space Station brought Russia in as a fourth partner in December 1993. Russia had developed its own space station through the Salyut and Mir programmes and brought much to the NASA-led station, but the prospect of using the Shuttle as a revenue-earning asset for commercial customers was gone.

Inglorious Retirement

In the early 1980s, NASA had looked at the prospect of privatising the Shuttle, offloading operations and technical support to a commercial organisation which would probably have consisted of the major companies that had built it. Several individuals raised interest in that prospect and lobbied Congressional leaders for support. That did have some appeal, the enormous cost of running the space programme was itself considered a burden on the economy while the revenue-earning telecommunications companies made significant profits. Surely there was a way to realise commercial potential, some said.

The reality that the Shuttle was unable to achieve annual flight rates to make it a financial success prompted questions about the value of the International Space Station and that led to significant delays in its development. For almost a year after the 1984 declaration by President Reagan that NASA was going to build the station, the programme had been in limbo until President Clinton ordered a complete rework of the design in mid-1993, several months before the Russians were brought on board. There is some traction to the argument that had it not been for the Russians, brought in to encourage better foreign policy relations, the station may well have been cancelled as Clinton had threatened to do before the deal was signed.

When the station was announced in 1984 hopes had been raised that it could begin to support a permanent crew by the early 1990s but expectations were dashed when it came very close to outright cancellation long before it was built. Not until 1998 would the first elements reach orbit followed two years later by the first residential crew, since when it has never been without occupation. Over a period of 15 years, the station was assembled into a facility with crew members from 28 countries contributing to research in orbit.

The Shuttle programme took another hit on February 1, 2003 when Columbia was destroyed re-entering the atmosphere after a long-duration science mission, completely dashing fresh hopes for a commercial potential. That single event proved transformative in that it pushed the agency to examine the future for this reusable transportation system and to re-evaluate the US manned space programme in its entirety. Not until July 26, 2005 did the Shuttle programme resume flight operations, with further restrictions on how it would be operated.

Because Columbia had flown an independent mission where it could not place its crew aboard the station to await a rescue flight, in future only under exceptional

circumstances would it be allowed to fly solo missions again without placing a crew aboard the space station where they could ride out any problems with the shuttle that launched them. NASA had to decide about the Shuttle's future and evaluated a comprehensive programme of modernisation and upgrades.

In 2004, President Bush announced that NASA was going to retire the Shuttle and use money instead to fund development of a small, crewed spacecraft looking much like Apollo and launched on a developed version of a shuttle booster with an upper stage, a combination named Ares I, to send crewmembers to the station and back. Called Orion, the new spacecraft would also be used with a much larger Ares V launcher to support a return to the Moon by 2020 for extended stays at scientific research stations on the lunar surface.

Before the Shuttle retired on July 21, 2011 after its 135th flight, that plan had changed significantly as NASA was finally set on a course in which its future would be largely dependent on commercial funding rather than by the American taxpayer. Old Space was about to give way to New Space and that would prove painful.

ABOVE • Before the loss of Challenger *in January 1986, over nearly five years the shuttle flew 25 missions, several of which carried commercial satellites into orbit. (NASA)*

BELOW • Long before the Space Shuttle conducted its last mission in 2011, thought had been given to operating it as a commercial business. Emblematic of that possibility, in fact this image was taken in November 1984 on the 14th Shuttle flight immediately after recovering two failed satellites. (NASA)

MY SPACE

The perils of bone loss and radiation

ABOVE • The opportunities for humans to live and work in space opened up greatly with the International Space Station, permanently occupied since 2000. (NASA)

Getting 'things' into space was never going to be easy. Transporting humans there is even harder. Are there limits to an untrained astronaut being able to take a trip into space? If so, what are they and how does that stand against the hopes of some and the dreams of many for taking the ultimate high-flying adventure?

Exposed to the weightlessness of space the human body goes through a staggering transformation and suffers with effects only increasing with time spent in orbit. In the beginning, when space flights kept astronauts in orbit for only a few days, there were faint signs of adverse effects. As those flights got longer signs began to appear which could have led physicians to conclude that there were limits to the amount of time people could remain in space without returning to Earth. Because of these uncertainties, NASA medical specialists and their Russian counterparts put astronauts and cosmonauts through a series of tests and experiments in which they became human guinea-pigs and studies of the physical effects were conducted on those who remained in space for up to a year or more.

The most immediate effect on arriving in space is a form of motion sickness which is caused by a loss of orientation in the inner ear, something which can affect people in aircraft

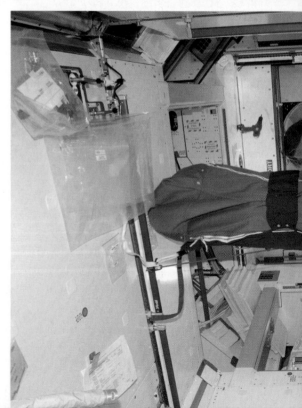

or at sea and is usually accompanied by vertigo, vomiting, headaches, and a general feeling of nausea. This is something that almost half of all who go into space feel, the effects usually lasting only a couple of days. The second Russian in space, Gherman Titov was the first to report these symptoms experienced during his August 1961 flight which lasted just over a day. In March 1969 NASA astronaut Russell Schweickart reported the same effects during the Apollo 9 flight, delaying a planned spacewalk. Many astronauts who experienced this effect remained silent for fear they would be removed from the active flight roster. But there is still no certain test than can precisely say who will get 'space sickness', as it is called.

The next most obvious effect is when a person falls asleep on their first night in space. With 18 sunsets and sunrises every 24 hours, isolating the visual effects of the continuously repeating periods of light and dark is essential. On Earth, when the normal day and night cycle is disrupted, people respond badly to night light, whether in the sleeper's room or from outside and there is little or no adaptation. In space, securing that day and night cycle is vital for good sleep but that is compromised by the continuously changing noise levels from fans and pumps circulating the air to prevent it becoming stagnant. In a weightless environment there is no convection and that causes air, and body odour to pool and remain undissipated unless physically driven away by mechanically driven circulation systems.

Having got over nausea and a bad night's sleep, personal hygiene brings its own unique problems. In the absence of gravity, human waste has to be physically removed by separating liquids and fluids into bags as they are released from the body, kneaded solids with a biocidal germicide to prevent bacterial growth until the contents can be returned through the atmosphere and burned up. Uncomfortable? Yes, and likely to produce aches and pains in all the wrong places as gravity fails to move solid waste naturally down the usual internal pipes! Which brings us to food, where a low-residue diet wins out in the short term but may be deficient in nutrition over extended periods.

In the early days of space flight, food came in tubes or freeze-dried items or in dehydrated form in bags rehydrated with saliva or water from a spigot attached to a pipe. Or dry, bite-size chunks coated with a gelatine to prevent crumbs floating away and becoming lodged in crevices. As flights got longer, weight became an issue. Each crew member was apportioned 1.27lb (0.58kg) of food per day. Food for a two-man crew would weigh more than 35lb (15.8kg) for a two-week flight. A complete menu of dehydrated juice freeze-dried and dehydrated foods and bite-size items made up three meals for each crewmember every day with planned meals repeated every four days.

ABOVE • Meals in space now mimic traditions on Earth, the first such gathering provided the three crewmembers of Skylab (1973-74) with a galley and food trays. (NASA)

LEFT • Studying human behaviour in space is key to future commercial opportunities but sleep takes a bit of getting used to where arms float upwards in microgravity. (NASA)

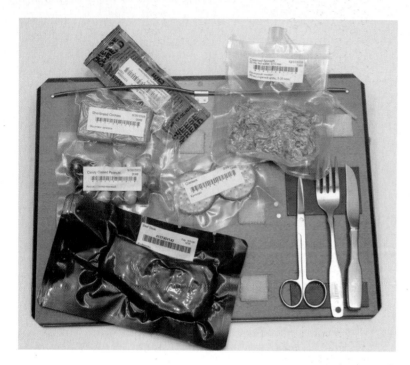

ABOVE • From the 1980s, food on the Shuttle was a step forward, with magnets to hold things down, scissors to open packages, and spoons for eating jelly and semi-liquid items. (NASA)

for an additional four weeks. The Skylab space station was designed with crew needs very high on the list of priorities for comfort and relaxation, considered essential for high work productivity and self-satisfaction with complex tasks taking up a large part of the working day. Specialists were brought in to advise design teams on interior colours.

Provision was made for a three-place dining 'table', essentially a fixed pedestal to which were attached three equally spaced warming trays. Foot restraints allowed each crewmember to remain in position to eat food prepared from special lockers on the walls of the station. With a dedicated galley area there was a wider selection, with items chosen placed within receptacles on the heated trays and warmed to a comfortable temperature. It broke new ground for space flight in providing a conduction heater and refrigerators storing ham, chilli, mashed potato, ice cream, steak, and asparagus – all carefully selected for their nutritional value but also to deliver a few tastes from home.

For Shuttle flights the challenges were different, up to seven crewmembers being provided with sufficient nourishment for up to two weeks with 74 different foods and 20 kinds of beverage. There was an approved list and a range of personally selected items to supplement the basic diet, all separately taste-tested by each astronaut before the flight and signed off by nutritionists and dieticians. Like Skylab, there was a galley and the age-old human habit of gathering together to eat food and exchange pleasantries was considered important for crew morale and integration, which always began immediately after crew selection a year or more before flight and frequently pursued on the ground as lasting friendships were forged.

The Shuttle galley had a water dispenser and an oven located in what was referred to as the middeck area directly beneath the flight deck, access between the two being made through a square hatch. The forced-air convection oven was another special design because in weightlessness there is no convection and heat will not rise of its own accord. With food heated in different size containers, a meal for four could be set up in about five minutes and reconstituted and heated in a further 30 minutes. And yes, there was a designated 'cook' per 'day' who prepared the meals for everyone!

During Apollo missions the variety of food available on missions grew as experience in the selection, preparation and packaging improved. Squeeze-tubes were replaced with spoon-bowls of food with a creamy texture that would stick to the spoon and not float away, and some food was provided in small cans with pull-off lids. With these new preparations, taste and smell increased, adding to the satisfaction of taking a meal while drifting around or restrained in one position. However, the new containers added weight and that was always a compromise but the storage space for a week's rations for one person would take up the volume of three shoeboxes.

The challenges of keeping people alive and fed for long periods on space stations is another issue altogether and this was first met with Skylab where teams of astronauts would remain in orbit for up to two months except, to get the most out of their stay, the final crew remained in space

RIGHT • A food table on the International Space Station gets just about as Earth-like as it ever will in a near-weightless environment, with many familiar eating utensils and a wide selection of drinks and appetising items on the menu. (NASA)

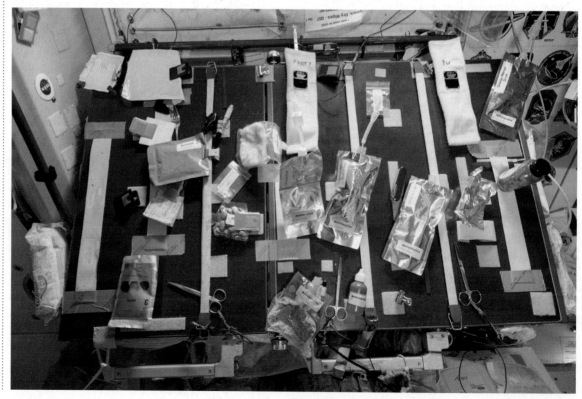

LEFT • Mealtimes can get a bit like camping out, literally under the stars, with some getting a heads-down approach to food on offer! (NASA)

The Shuttle programme lasted 30 years and included 135 launches, during which time there were some changes and improvements to the food, but there also added challenges when some flights got longer, and greater quantities were required. Initially, food was not much different from the Skylab period, but that changed and rigid containers were replaced with lighter and more flexible material and a special compactor was added to reduce the volume of the waste to which the redesigned packaging was also added.

The International Space Station benefitted from the development of food storage, preparation and consumption from the Skylab and Shuttle programmes, but the quantities were very much greater, with crew levels between six and ten people at any one time. The freezers, refrigerators, and general storage facilities were greater, and food is usually delivered every three months to satisfy requirements for the next quarter-year period.

But visitors to the station benefit from the preparation of a wider selection of food including tortillas which are perfect for weightlessness and burritos, hamburgers and peanut butter and jelly sandwiches, most of which is prepared at the Space Food Systems Laboratory at NASA's Johnson Space Center, Houston, Texas. And there is always the occasional supply of fresh fruit and vegetables, brought to the station by the unmanned commercial cargo vessels that keep it replenished.

Fitness Regimes

But health and wellbeing in space is hard work, as the weightless environment brings penalties for the lack of gravity. Early in the manned flight programmes conducted by the Russians and the Americans it was noted that bones begin to lose their strength as calcium levels are reduced and demineralisation leaves them much weaker. Moreover, without gravity, the human body automatically adjusts the levels of water as it would on the surface of the Earth. Evolution has produced a system where the natural tendency for water to pool in the lower parts is compensated by an equally natural process to move fluid back up.

In weightlessness, water remains uniformly throughout the body while the natural processes add to the fluid levels in the torso and upper parts, including the head. Which is why astronauts get puffy faces and headaches as the pressure from all that fluid increases. There is just no way to adjust that naturally but much like the loss of calcium in the bones, rigorous exercise can help strengthen languid muscles now unnecessary to move massive objects around and which also deteriorate quickly over time.

But there is another potentially more sinister consequence of space flight, the loss of red blood cells which are produced in bone marrow along with white cells. Approximately two million red blood cells are lost every second to an Earth-dweller but in space that rate increases by 54% and they are not replaced. The effects are akin to anaemia where insufficient red cells induce fatigue and a general sense of malaise with impacts on the heart and on brain function. It is thought that the shift in body fluids due to weightlessness may be the cause. Initially, it was thought that red cell counts were an effect of immediately experiencing weightlessness, but it is now known that the levels go down and stay there and will not return to normal for up to a year after that person returns to Earth.

BELOW • Fresh chilli peppers grown from seed and farmed on the International Space Station add variety to the menu. (NASA)

RIGHT • Astronaut
Sunita Williams
keeps fit on the
Cycle Ergometer and
Vibration Isolation
System (CEVIS) bicycle
providing a low-
impact, high-cardio
workout. (NASA)

Key to reducing the debilitating effects of spaceflight is exercise and the amount is measured by the physiology of the space traveller — no two people are the same and, as on Earth, there is no fixed regime applicable to everyone. Each human is unique and constant measurement of their physical condition is an important part of maintaining health and countering the effects of weightlessness. Exercise machines have been uplifted to the space station so that crewmembers can spend several hours a day working out to reduce these effects and that will always be an essential requirement for people living in space for more than a few weeks.

One effect for which there is little mitigation is the physical reaction to radiation, one of the greatest threats to human life and healthy bodies. The surface of the Earth is protected from energetic particles from the Sun and cosmic rays from violent events in the galaxy by the magnetosphere which acts as a shield, but which also contains trapped particles. All forms of ionised radiation consist of atoms stripped of their electrons as they travel close to the speed of light, and all are very harmful to the body.

Ionising radiation is more difficult to avoid because it can move through physical structures and alter them as it passes, leaving significant damage and triggering additional damage from secondary particles thus created. The magnetosphere is formed as particles from the Sun, known as the solar wind, hit the Earth's magnetic field and compress it into a bullet shape flattened to within 35,000 miles (56,315km) on the side facing the Sun, to a tail which extends far beyond the orbit on the Moon. Inside the magnetosphere radiation particles which can be harmful to humans are trapped in magnetic fields known as the Van Allen belts, doughnut shaped structures with Earth at the centre. These extend from about 620 miles (1,000km) to approximately 37,300 miles (60,000km).

Only 24 Apollo astronauts who left Earth orbit for the Moon have gone through this region and 12 of those landed on its surface. No other astronaut, cosmonaut or space traveller has ever left Earth orbit and almost all of those have never been further than 500 miles (804km) from its surface. Considering that the Earth has a diameter of 8,000 miles (12,872km) that is not very far at all, but it is sufficiently high above the atmosphere for a space traveller not to get pulled back down before completing at least one orbit.

The challenges for human health faced by astronauts returning to the Moon are immense, all the greater for those taking a flight to Mars where a round trip plus stay time could last two years. The Moon has only one-sixth the gravity of Earth and while that alleviates some of the physiological problems experienced with weightlessness, Mars has one-third the gravity of the Earth and this will greatly reduce problems with working and living on its surface. But anywhere beyond the magnetosphere around Earth, space travellers will have the radiation challenge and engineers will have to provide storm-shelters for protection against excessive high energy levels from the Sun which occur at infrequent intervals.

Meanwhile, current levels of acceptable risk dictate that due to the danger of endometrial, ovarian and breast cancer, female astronauts are only allowed half as much accumulated time on space flights as their male colleagues. Both standards and protection from harmful radiation on missions to Mars

BELOW • Keeping
regular checks on
health and fitness
is vital to maintain
physical wellbeing.
Danish astronaut
Andreas Mogensen
takes blood samples.
(NASA)

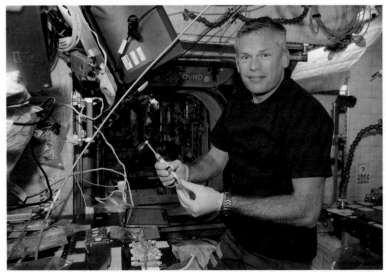

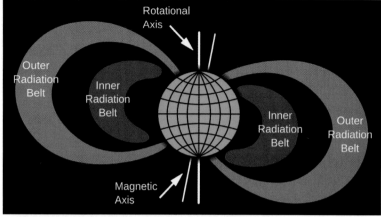

will have to be provided to avoid severe limitations on groups who can make the trip. Which is why understanding the very different female physiology is so important in crew selection and future mission roles for women on long duration flights.

In the early years of the American space programme, women were not considered to possess the necessary mental abilities, aptitudes, or qualifications for becoming astronaut, an outrageous assumption considering the number already working on specialised activities for NASA in mathematics, physics, electronics and engineering design and development. At the time, women were excluded from positions as military air crew despite many having pilot's licences after serving in World War Two flying transport aircraft and delivering bombers from factories in Canada across the Atlantic Ocean to the UK. But this was a time when only 25% of women worked, when there was no career path for them in the military, within a social norm in which married women had to get their husband's written permission to open a bank account, even to buy property or large household goods.

Woman for Space

The first NASA astronaut selection was completed with the announcement on April 9, 1959 that seven men had been chosen for the Mercury programme, a group hereinafter known as the 'Mercury 7'. The Russians began their selection programme later that year and appointed 20 pilots to train as cosmonauts, one of whom was Yuri Gagarin who made the first manned space flight on April 12, 1961. At that time, neither country had plans to recruit women for space but in the United States a pioneer in screening applicants for astronauts had a different idea. A qualified surgeon and a notable physiologist involved in developing tests for male astronauts, William Randolph Lovelace II wanted to know how women would respond to the same selection process.

Lovelace had been involved in selection of the Mercury 7 and had funds and resources to start an unofficial

programme to offer suitable women the chance to follow precisely the same programme that the male applicants had gone through and to submit to the same physical examinations and tests. There was no realistic possibility that women would be selected, but he wanted to shadow the male astronauts with this duplicate process for medical and physiological research. While the majority of men refused to consider women for the programme, Lovelace and Brigadier General Donald Flickinger of the US Air Force believed that for several reasons women would be better suited for space travel than men, purely on practical and engineering grounds that they generally weigh less, would not require muscular strength in weightlessness and the fact that for a given task they consume less oxygen than their male counterparts, cutting weight requirements.

Thirteen of 19 'Lovelace women' volunteers selected passed scrutiny, a 69% success rate compared to 56% of male applicants in the official Mercury process. Moreover, in running what was called the Women in Space Earliest (WISE) programme, Lovelace discovered that in the majority of parallel tests women generally scored higher marks than men. He found that they were more resilient in physical screening and particularly high in mental agility, decision-making and in 'responsive reactions' to crises put in their way during training sessions.

The acronym WISE was a parody of the earlier US Air Force programme named 'Man in Space Soonest' (MISS) before NASA took it over as Mercury. To show equality, the women self-named themselves the 'Mercury 13' but as interest grew in Congressional hearings, when asked whether he approved of women being selected for space flight, John Glenn (appointed as one of the Mercury 7 without the

ABOVE • Earth's magnetic fields trap charged particles in what are known as the Van Allen radiation belts through which astronauts must pass to escape low Earth orbit, posing potential health hazards. (Author's archive)

FAR LEFT • NASA astronaut Kjell Lindgren works out on an Advanced Resistive Exercise Device (ARED) to offset the adverse effects of weightlessness. (NASA)

BELOW • In the late 1950s, women wanted to train as astronauts but were not allowed to apply while 13 were selected for a parallel programme to NASA's Mercury project run by Dr Lovelace. Left to right: Gene Jessen, Wally Funk, Jerrie Cob, Jerri Truhill, Sarah Ratley, Myrtle Cagle, and Bernice Steadman from that group. (Lovelace Clinic)

requisite college degree) replied that he thought a woman's place was in the kitchen and that they should remain there.

The WISE programme ended in mid-November 1959 and Flickinger asked Lovelace to take it over, which he did, providing some of the most erudite benchmark research into female physiology, exploring all the many and varied reasons why studying women in the space environment would add greatly to the benefits for general health and medical care for women on Earth. The final put-down came when President Lyndon Johnson prevented a letter written by his executive assistant Liz Carpenter being delivered to NASA boss James Webb appealing for women to be admitted, scrawling across the bottom "Let's stop this now – File!"

On hearing NASA's plans for its manned Apollo programme, Russia executed a propaganda coup when it flew Valentina Tereshkova into space in Vostok 6 on June 16, 1963, a dual flight while Vostok 5 carrying Valeri Bykosvsky was still in space. The two never came closer to each other than 3.1 miles (5km). Tereshkova returned to Earth after two days 22hrs 50min, her one flight exceeding the accumulated hours in space of the four Mercury orbital missions to that date. She had been one of five in the group of female cosmonauts selected on February 16, 1962 but it would be almost 20 years before another Russian woman made it into space.

Realising that they had launched only one women thus far, focus turned to beating the Americans by launching their second female cosmonaut. Svetlana Savitskaya was selected as one of nine on June 30, 1980, after NASA announced that it was planning to recruit a new cadre of astronauts which included women. Savitskaya was launched in a Soyuz spacecraft on August 19, 1982 and spent time on the Salyut 7 space station, returning on August 27 after almost eight days. She recalled later the banter that accompanied her arrival on Salyut 7 when a cosmonaut already on board presented her with an apron and told her to "Get on with it". It was the first mixed crew in space.

By this time, the changes in America's societal norms had broken the barriers that previously prevented NASA from recruiting female astronauts which, given the cramped quarters on the early manned vehicles, had practical difficulties. But

with the Shuttle programme, the 1980s opened opportunities for America to also fly mixed crews. Unlike the Mercury, Gemini, and Apollo spacecraft, the Shuttle provided spacious living quarters and privacy impossible with earlier vehicles.

Shuttle crews would consist of categories for pilot and mission specialist, all of whom had to be professional astronauts. A new category of payload specialist provided an opportunity for people to fly in space managing specific science experiments or payloads but who were trained specifically for their unique role. In this way, specialists who had special value for particular missions could bypass the lengthy process of applying for the role of professional astronaut and contribute specialised skills. It was within the mission specialist category that NASA first wanted to bring women into the astronaut corps. Responding to a recruitment campaign in 1977, Sally Ride was one of six women and 29 men selected on January 16, 1978, of whom 15 were pilots and 30 were mission specialists.

As NASA's first female astronaut in space, Sally Ride made her first flight on the seventh Shuttle mission launched on June 18, 1983. She operated the remote manipulator arm to deploy and then retrieve a package of science instruments which were returned to Earth in the payload bay. Three months later she met with Savitskaya on her own initiative and the two spent six hours conversing. Savitskaya would make her second flight aboard a Soyuz spacecraft to the Salyut 7 station on July 17, 1984 where she became the first female astronaut to conduct a spacewalk during which she conducted an external welding task.

Launched on October 5, 1984, Ride made her second flight on the 13th Shuttle mission, accompanied by Kathryn Sullivan who was making her first flight and also the first in which two female astronauts were in space together. Ride was assigned a third Shuttle flight, but the loss of Challenger on January 28, 1986 dramatically changed those plans. She would never fly into space again but played a key role in the Challenger investigation without ever revealing that it was she who had warned about the effect of extreme cold on the O-rings that failed and caused the loss of the Shuttle. To protect her anonymity, the information was passed to fellow commission member Richard Feynman, inappropriately remembered as the originator of that revelation.

While no female Russian cosmonaut flew into space for more than 10 years after the second flight of Savitskaya in 1984, NASA continued to assign women to Shuttle flights, but the 15th female astronaut was Britain's Helen Sharman, launched aboard a Russian Soyuz spacecraft on May 18, 1991 paid for on the direct intervention of Mikhail Gorbachev. The opportunity arose under Project Juno, a private enterprise backed by commercial and media sponsors to put the first British astronaut aboard the Mir station, the 'commercial' flight aboard a Soyuz spacecraft costed by the Russians at

£7m. The venture was an effort to enhance British/Soviet relations and to extend goodwill to the Russian people.

The selection process for a suitable astronaut attracted more than 13,000 applicants and a selection board chose four candidates, three of which were male and one female, Helen Sharman. On November 25, 1989 Helen was selected along with Major Timothy Mace for special training at Russia's Star City outside Moscow, the equivalent of NASA's Johnson Space Center. There they received tutorials in spacecraft systems, mission operations, safety procedures and what to do in the event of an abort which would jettison the capsule from the rocket and bring them down to Earth. They had to become fluent in the Russian language and undergo fitness tests with measured responses to psychological and temperamental stress.

After 18 months in the Soviet Union, they were declared ready for flight, but Project Juno was in trouble, promised funds failing to arrive and sponsorship falling away. The media became less interested and judged that the rewards for their publicity coverage were not sufficient to justify the cost. The project faced cancellation until the UK government stepped in and, while unwilling to spend taxpayer's money on what was considered a private and commercial venture, approached the Russian government to begin a process in which Gorbachev intervened and saved the day.

Money was found through the Moscow Narody Bank, which had strong ties with Morgan Grenfell. Intervention at a government level reinforced the value this project seemed to those seeking a stronger tie with Western democracies in its effort to transition from communism. But these events occurred on the cusp of the collapse of the Soviet state and popular interest in the unique opportunity provided by Juno fell away, and a compromise cut several science experiments the British astronaut was to have conducted. Helen Sharman was chosen over Timothy Mace and flew on Soyuz TM-12 with Russian cosmonauts Anatoly Artsebarsky and Sergei Krikalev in a flight launched on May 18, 1991 lasting seven days 21hrs 13min.

Sharman returned on May 26, 1991 aboard Soyuz TM-11, which had delivered a resident crew to the Mir station along with a Japanese reporter, Toyohiro Akiyama, five months previously. His flight on December 2, 1990 had been sponsored by the Tokyo Broadcasting System (TBS) which paid more than $25m for training at Star City and the seven days 21hr 54min flight aboard Mir. Akiyama's trip was the first fully commercial space flight, providing Russia with foreign currency.

A new era had dawned. Paradoxically, the Russians had been the first to make money in a commercial deal for their storied space programme. Others were watching and had similar ideas, not all of them looking to do it the conventional way.

BRANSON'S BIG GAMBLE

Selling rocket-rides to tourists

On September 25, 2004, British entrepreneur, adventurer, and highly successful businessman Richard Branson announced that he was heading for space. The man that had made his fortune in record labels, airlines and railway companies signed with Scale Composites funded by Microsoft co-founder Paul Allen and aerospace engineer Burt Rutan. Branson wanted to open up space as a new and exciting business venture. Very few saw that coming but it gave name to a new company, Virgin Galactic.

On offer were rides to the edge of space for rich thrill-seekers looking for the ultimate, out-of-this-world experience providing several minutes of weightlessness. There was precedent for thinking of space as the next bucket-list destination and Branson was the first to propose what he believed to be a quicker and cheaper way of giving ordinary people the experience of a lifetime by making such visits easier than the evolving trend for riding big rockets.

By the late 1990s, Russia was capitalising on its post-Soviet, pro-capitalist approach to business with commercial flights to the International Space Station, brokered by Space Adventures founded by Eric C Anderson in 1998 through offices in Virginia and Moscow. As a major partner in building elements for the ISS and launching them into space, Russia had a justifiable claim on managing commercial flights, much to the initial annoyance of NASA who saw such activities as inhibiting scientific work on board the station.

Nevertheless, in April 2001, businessman Dennis Tito paid around $20m for a flight to the ISS, the total trip lasting almost eight days. To date, the Russians have sent nine tourists to the station on eight separate missions, fee-paying 'passengers' riding along with professional cosmonauts heading to the ISS for crew duties. The most recent in December 2021 was unique in that it was a purely commercial flight without delivering any professional astronauts as crewmembers. Since that date, the opportunities for riding Russian Soyuz spacecraft to space have suffered as a consequence of the war in Ukraine.

But it was the emerging opportunities for attracting fee-paying customers, ready and willing to pay for such trips that prompted Branson to build Virgin Galactic based on the technology available at Scaled Composites, the enabling key to Branson's dream. The origins to that lie within the Ansari X Prize which was set up in 1996 as the X-Prize but renamed in 2004 after a major donation from entrepreneurs Anousheh and Amir Ansari. It offered $10m to the first private organisation to make two

company for designing and testing experimental aircraft. Innovative and radical in his approach to aircraft design and operation, Rutan designed SpaceShipOne to reach more than 62 miles (100km) which is the internationally recognised outer edge of the atmosphere above which space is regarded to begin. Although in reality there is only a gradually diminishing presence of atmospheric molecules, the 100km transition point is an arbitrary number above which a complete orbit of the Earth is possible, but only for highly elliptical paths, and for that reason it is standard for internationally accepted records. The lowest altitude for a circular orbit is 77.67 miles (125km).

The 100km height above the Earth's mean sea level is known as the Kármán line in recognition of the work of the Hungarian-born engineer, Theodore con Kármán. He conducted work both in aeronautics and astronautics to find a point at which the atmosphere ends and space begins. It is the line recognised by the Fédération Aéronautique Internationale (FAI) which sets the standards and the conditions for recognising records broken for both aviation and space activities. It was for this reason that the X Prize recognised 100km as the qualifying altitude for recognition as a flight into space.

Different values have been proposed for different reasons. The Apollo programme accepted an altitude of

trips of a manned vehicle into space and back successfully within two weeks.

Twenty-six teams participated but only one could win and that was Tier 1, the attempt by Scale Composites using a combination mother-plane/spaceplane which it achieved on October 4, 2004, the anniversary of the launch of the world's first artificial Earth satellite. It was also the perfect stimulation for Branson's idea of running a space tourism company on the technology demonstrated by the little spaceplane, named SpaceShipOne which was carried high into the atmosphere by the WhiteKnight carrier-plane where it was released for a rocket-powered flight to the edge of space and back.

The story began in 1982, the year after the Shuttle first flew into orbit, when Burt Rutan formed his technology

ABOVE LEFT •
Accompanied by
Russian cosmonauts,
Dennis Tito spent just
over a week in space
integrated with full-
time crew activities.
(NASA)

ABOVE RIGHT •
South African Mark
Shuttleworth became
the second self-funded
space tourist when
he flew in a Russian
Soyuz, spending a
week aboard the
International Space
Station. (NASA)

BELOW • The X
Prize stimulated
Burt Rutan to build
a vehicle which could
fly to the edge of
space twice within
a two-week period.
Left to right: FAA
chief Marion Blakely,
pilot Mike Melvill,
Richard Branson, Burt
Rutan, pilot Brian
Binney and Paul Allen.
(Don Ramey Logan)

83.3 miles (134km) as the height above the Earth's mean sea surface where aerodynamic effects are first felt on entering the atmosphere. This was also accepted for Shuttle recovery, albeit where there is much less energy returning from orbit at 17,500mph (28,157kph) rather than around 25,000mph (40,225kph) returning from deep space. NASA's definition is based on the point at which there is sufficient aerodynamic friction to measure a retardation level of 0.05g at which a returning vehicle encounters the atmosphere.

Conversely, the US Air Force accepts an altitude of 50 miles (80.45km) and since 2005 NASA has reverted to this measure, whereas previously it accepted the 100km altitude, for formal recognition of pilots having transited from the 'atmosphere' into 'space', particularly flights made by USAF and NASA pilots in the rocket-powered X-15 programme of 1959-1968. The 50 miles line is also accepted by the US Federal Aviation Administration. Either way, the X Prize satisfied these various definitions and, therefore, Branson's claim to carry people into space is universally accepted as valid for a very special place in an equally prestigious record book.

In the Company of Spaceships

When Branson set up Virgin Galactic in 2004, he had a clear vision for how he would market the capabilities of SpaceShipOne but in a developed form for an exclusive market niche. There would be an introductory offer of reservations with a deposit based on a reduced fee for a limited period while Scaled Composites developed a suitable successor for routine flight operations. Some

specialists in aviation and space projects questioned the attraction of making a flight inside the spaceplane carried into the high atmosphere by a mother-plane before a rocket-ride to just beyond the Kármán before slowly falling back down to a conventional landing after a few minutes of weightlessness.

The effect could be experienced for a lot less money than a ride in a spaceplane. Astronauts train for weightlessness in a parabolic flight in an aircraft sufficiently large to carry several crew inside what passes for a hollowed-out airliner. The effect is created in a parabolic flight where the pilot pulls the nose up and as it climbs the engines are throttled back. When the aircraft reaches the high point in its trajectory and begins a free fall, occupants experience weightlessness for around 22 seconds. As the aircraft falls it pitches down 42° before gradually levelling off, exerting a force of 1.8g during recovery and a slow descent, either to a repeat of that cycle or a return home.

The Branson ticket offered up to five minutes of zero-g experience before decelerating to a low flying speed and landing on a conventional runway. But the total flight time from taking off attached to the underside of the mother-plane is about 2.5 hours and the fastest speed achieved would be 2,000mph (3,218kph), considerably less than the speed required to reach orbit, about 17,500mph (28,157kph). The ticket prices initially offered a subsidised ride for $250,000 but that cost was expected to reduce as flights progressed, the development of a fully operational system envisaged by Branson being up to 400 flights a year. Branson expected that repeat flights would help increase the flight rate and bring costs down, allowing a reduction in price and attracting more customers.

Major publicity surrounded the launch of this plan, with Branson expressing confidence in a system which Rutan had been working on since 1994, the X Prize being the fulfilment of a dream that took three years to reach fruition from an initial start on the design and assembly of SpaceShipOne. Under Branson's Virgin Galactic, Tier 1b development would help finance the development of a passenger-carrying successor. As marketed, the opportunity to see the Earth from space and to experience weightlessness carried less risk than orbital flight begun riding a ballistic rocket requiring 60 times as much energy and with a requirement for a heat shield to protect the vehicle on returning through the atmosphere.

Expectations were high that the first commercial flights could begin as early as 2010 with SpaceShipTwo and a new carrier-plane, WhiteKnightTwo ready by 2008. The Spaceship Company (TSC) was formed by Branson and

Rutan in 2005 to provide these vehicles for Virgin Galactic to market launch services, delivering five SpaceShipTwo and two WhiteKnightTwo vehicles for passenger-carrying operations. All of which challenged the Federal Aviation Administration (FAA) to set up a regulatory framework in which these activities could take place and provide licences for commercial flights. This was the outline plan under which Virgin Galactic sought sustainable funding to develop a long-term programme.

Initially, Virgin Galactic claimed an investment of $100m and in 2010 Aabar Investments, the sovereign wealth fund of Abu Dhabi, acquired 31.8% of Virgin Galactic at a reported injection of $280m. In doing so it bought the rights for flights also to be conducted from the UAE's capital city as part of an initiative to build investment in science and technology as it sought ways to diversify from oil. The key to the UAE embarking along this path lay in Branson's claim that the Virgin Galactic venture would eventually embrace the launching of small satellites into space, for which in 2011 Aabar planned to inject $110m also covering a small spaceport in the UAE and raising its equity to almost 38%.

The fundamental technology was appropriate for transferring basic principles from the vehicles developed for commercial passenger flights to a launch system which could fill a market niche for sending small payloads into orbit. As financial markets crashed and the world plunged into an economic recession, costs of sending satellites into orbit became a pivotal issue in many small-scale development projects. The cost of sending things into space had brought an end to NASA's Apollo programme and while those expenditures represented the high end of space activity, at the entrant end it was similarly stressed searching for ways to find cheaper access to weightlessness and its value for scientific research programmes.

Over time these plans would grow, albeit slowly, with the LauncherOne concept emerging as a commercial proposition. The essential technology was the enabling catalyst for these ambitious goals but the practical application to a meaningful business required facilities and launch sites, for which development of a major complex in the Mojave Desert was the preferred choice, emerging in 2004 as the first FAA-certified spaceport. It was to be capable of both horizontal and vertical launches, carried by carrier-planes and rockets, respectively.

The sheer existence of this facility, matched with the X Prize/Ansari achievement crossing the Kármán line, and the commercial plans of Richard Branson, brought new expectations of public access to space and thrill-rides for tourists ticking off the ultimate item on the traveller's bucket-list. Behind the deals and the signing of contracts to much media fanfare, lobby groups and low-level focus groups found enough potential among those with sufficient disposable income to justify financial support and within several weeks of the announcement, people from around the world had placed deposits for seats on SpaceShipTwo.

In publicly announcing these plans, Branson began a wide-ranging search for similar opportunities which could be provided through commercial deals outside the constraints of politically-managed, government programmes and it sparked interest from people who had previously never expressed a desire to reach space. And the prospect of developing small satellites for relatively low-cost flight also excited research scientists and organisations around the world, sparking interest among those seeking ways to fly such payloads and those already manufacturing small satellites for launch on small rockets. The dream had great potential but would take far longer to achieve than first anticipated, and cost much more than expected.

ABOVE • Theodore von Kármán and assistants work out flight paths, his seminal work on the transition line between the atmosphere and space forming the basis for record-breaking flights. (USAF)

BELOW • SpaceShipOne and the WhiteKnight carrier-plane established an operational benchmark on which Richard Branson would base his business through Virgin Galactic. (Don Ramey Logan)

REACHING FOR THE KARMAN LINE

Flight in the weightless void

The incentive for building the world's first passenger-carrying vehicle capable of routine flights to the edge of space and back grew out of the ambitions of talented dreamers with engineering expertise and a creative imagination. Since aviation pioneers first tried installing rocket motors in conventional aircraft during the 1920s, inventors and entrepreneurs had wrestled with performance and reliability to fly faster, higher, and across the transition boundary into space. New materials, high-speed flight research, hypersonic wind tunnels, and experience with rocket motors had already shown that manned vehicles could be made to do that, but it would take a lot of work to turn that possibility into commercial reality.

The background to post-war, rocket-powered aircraft includes research flights conducted by the Bell X-1 series in the late 1940s and 1950s in which Mach 1 was exceeded, transonic and supersonic aerodynamics providing a base on which new generations of combat aircraft emerged. The X-1 series was powered by a motor designed by James H Wyld, a founding member of Rocket Motors Incorporated (RMI), and designated 6000C4, designating its thrust in pounds-force and the four combustion chambers it had. Each could be started independently offering 25% of 6,000lb (26,689N) of thrust and capable of incrementally increasing overall thrust to maximum as each was lit in succession.

To conduct research at and above Mach 3, Bell produced the troublesome X-2 powered by a Curtiss-Wright XLR25 with a total maximum thrust of 15,000lb (66,720N) from a dual chamber motor. It did achieve Mach 3 but at the expense of the pilot, Milburn Apt who was killed on September 27, 1956. Further research into jet engines and rocket motors resulted in a hybrid propulsion system for the Republic XF-91 Thunderceptor, first flown on May 4, 1949. It was powered by a General Electric J47 jet engine and by a four-chamber 6000C4 rocket motor, badged as the XLR11-RM9, for supplementary boost during take-off and when extra speed was required in flight. It was not successful and only two experimental prototypes were built.

The next most effective marriage of a rocket motor to an aircraft was the hypersonic North American X-15, powered by a single, throttleable Reaction Motors XLR99-RM-2 with a maximum thrust of 57,000lb (253kN). Like its predecessors, it was a conventional design with fuel and oxidiser stored in separate tanks and brought together in a combustion chamber. Three X-15s made a combined 199 powered flights between 1959 and 1968 but much had been learned about marrying a rocket motor to an airframe capable of operating at the edge of space, eight X-15 pilots awarded 'astronaut' wings for having exceeded 50 miles (80km) in altitude. A further tranche of contemporary spaceplanes was considered, most notably the Boeing Dyna-Soar (for 'dynamic soaring'), relegated to a test programme in which it was to be known as the X-20. None were ever flown.

Launched on an adapted Titan rocket, Dyna-Soar presaged the era of reusable space vehicles in search of

BELOW • An underside shot of the joined configuration with relative location of the two turbofan engines. (Scaled Composites)

low-cost space transportation and as noted previously, that led to the Space Shuttle which made its first flight in 1981. Considered by some an aircraft with rocket motors and by others as a spacecraft with wings, the Shuttle was a true transition vehicle which conducted 135 launches, the last in 2011. Ardent supporters of the spaceplane concept persisted in their belief that the Shuttle was but the first in what they imagined to be a future filled with winged aero-vehicles ferrying crew and passengers between the surface of the Earth and habitats in low Earth orbit.

Which is where the Ansari X Prize received its motivation and enthusiasm for a privately funded programme capable of stimulating entrepreneurs and creative aeronautical engineers. It worked and led to Richard Branson's support as a travel agent for technologists in the United States which would provide the means to market services to a wide range of people. Having publicly stated that he wanted to open space travel to everyone, Branson's self-declared objective was to make it such a common occurrence that anyone with disposable funds for a first-class, trans-Atlantic air ticket would be in a position to book a seat with Virgin Galactic.

Marketed as a portent of new and exciting possibilities, the Tier 1 programme that won the prize for a repeat trip to the edge of space within two weeks created the hardware and the operational capability to begin adaptation of an experimental spaceplane to a commercially viable passenger vehicle. It had to be scalable in operational frequency, safe and reliable, cost effective and with high levels of safety. The objective was simple but getting there would be a lot harder than anticipated and it was not at all certain at the time that Richard Branson had fully understood the challenges, or the amount of money it would take to realise his dream.

Spaceshipone

Scaled Composites' Model 316 SpaceShipOne concept had a unique specification as the first privately funded vehicle built to fly to the edge of space and return. But it was not designed for commercial use, or for carrying fee-paying customers. That would come with SpaceShipTwo as a result of the marketing deal with Richard Branson. SpaceShipOne was built to compete for the X Prize and for that alone. However, designed by Burt Rutan, it would almost inevitably have the

capability to evolve into a successor capable of carrying passengers.

The specification required that it carry three people including the pilot, that it would be carried into the air by a carrier-plane and that it would be propelled by a rocket motor to raise its altitude from 9.3 miles (15km) to more than 62 miles (100km) and by exceeding the Kármán line qualify for having reached space. Descending back down through the atmosphere it had to reduce speed created by kinetic energy as it got closer to the surface and be aerodynamically stable throughout both powered and gliding flight to a conventional landing.

Rutan and his small team of design engineers and technicians examined a wide range of possible alternatives and settled on the carrier-plane concept for lifting a rocket-powered vehicle into space. Simplicity and reliability would underpin safety and ensure a low risk of accident, but the vehicle had to operate in two distinct regimes, supersonic for going uphill to the edge of space and aerodynamically stable and controllable subsonic flight back down to the ground. Supersonic flight was always going to be easier than the controlled descent where the aerodynamic design would allow the vehicle to dump excess energy and remain stable as it reduced speed for a landing.

Initial configurations had a shuttlecock design with stabilising fins at the rear to 'feather' the aerodynamic profile as it descended, but that compromised the energy management profile and would have required mid-air

ABOVE • One of the first rocket powered aircraft projects, the Opel RAK. 1 was designed by Alexander Lippisch, purchased by Fritz Opel, and equipped with two solid propellant rockets provided by Friedrich Sander for its first flight on September 30, 1929. Depicted here is a replica at the Deutsche Segelflugmuseum in Wasserkuppe. (Bergfalke2/ Wikipedia)

BELOW • Bearing the US civil registration N328KF (indicating the altitude goal in feet), the SpaceShipOne is displayed here in fully feathered configuration. (National Aeronautics and Space Museum)

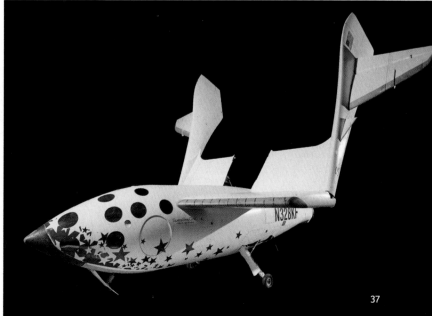

RIGHT • Just 18 years after Opel RAK. 1, Bell built the X-1 rocket-powered aircraft, the first to officially exceed the speed of sound in level flight on October 14, 1947. Such was the progress that less than 10 years later the Russians would put a man in space and 12 years after that three Americans would land on the Moon. (NACA)

retrieval and recovery. This was judged unacceptable and contravened the law of simplicity in aeronautical design which is that added requirements multiply the potential for something to go wrong; hence the acronym KISS (Keep It Simple Stupid!), first entering the lexicon of aviation design through the US Navy in the 1960s when land-based aircraft were adapted for shipboard use and made more complex because of those additional requirements.

The aerodynamic form of SpaceShipOne conforms to that of a shuttlecock because that is the most stable in-flight shape, largely irrespective of orientation and that factor would save it on at least one test re-entry. The structure was fabricated from a graphite epoxy composite material and the cigar-shaped fuselage was 5ft (1.5m) in diameter with a crew cabin at the front and a compartment for the rocket motor at the rear together with associated propellant tank and a fuel casing. Thermal considerations determined the location of the rocket motor, its placement controlled by the centre of gravity and the centre of pressure.

Lift was provided by a high set, stub wing protruding from each side of the mid-fuselage with a span of 16ft (4.87m) and a chord of 9.8ft (2.98m). The 1.6:1 aspect ratio

wing provided a surface area of 160ft^2 (14.86m^2). A large vertical tail boom was attached to each wing tip with horizontal stabilisers projecting outward presenting a total span of 26.91ft (8.2m). The tail booms were fixed to the aft section of the wings and the complete assembly pivoted up 70° to 'feather' the vehicle and create a high drag for reducing the descent rate, albeit incurring a deceleration of more than 5g.

The pressurised compartment in the forward fuselage was designed for a single pilot in front and two passengers behind, seated side-by-side. The proclaimed shirtsleeve environment was just that, the only occasion since the flight of Russia's Voskhod 1, launched on October 12, 1964, in which a vehicle capable of reaching space did not provide emergency life support in the event of catastrophic de-pressurisation. Every Russian and US manned space vehicle before and since has provided crewmembers with a suit for emergencies, including a back-up supply of oxygen. The design philosophy for SpaceShipOne incorporated dual seats and 16 windows, all of which had two panes. The occupants could access the interior either through a removable nose section or a hatch below the rearmost windows on the port side of the fuselage.

Control in roll, pitch, and yaw inside the sensible atmosphere – defined as a level of atmospheric pressure such that it resists a body passing through it – was by aerodynamic controls such as might be found on any small aircraft, with upper and lower rudders and elevons on the wings. Control where the density of the atmosphere was too low for aerodynamic reaction was by two sets of redundant thrusters, 12 in all with two at each wing tip for roll, two above and below the nose for pitch, and two each side of the fuselage for yaw.

The landing gear consisted of two widely spaced main wheel assemblies and a single nose skid released by springs and deployed by gravity. They were not retractable in flight and there was no pneumatic or hydraulic system to drive deployment should there be a hang-up after activation of the loaded springs.

BELOW • Conducting research into supersonic and hypersonic flight, the North American X-15 provided valuable research information and became the first winged vehicle to cross the Kármán line, 100km in altitude above Earth. (NASA)

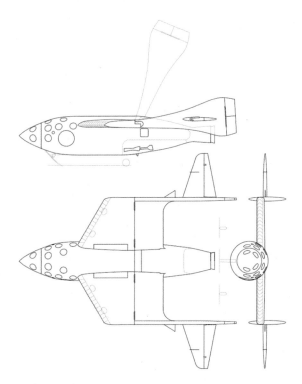

Propulsion was provided by a single rocket motor for one operation per flight. Developed by SpaceDev which had been formed in 1997 by Jim Benson, the company acquired all the patents and rights of the American Rocket Company (AMROC) and from that data base they designed the hybrid rocket motor for SpaceShipOne. Rutan and Benson chose a rubberised fuel, or HTPB (hydroxyl-terminated polybutadiene) to burn with a liquid nitrous oxide oxidiser as it propelled the vehicle through the upper atmosphere.

An important factor in the choice of rocket motor was the thrust/weight ratio: the heavier it was the more thrust required to propel the loaded vehicle to the Kármán line and beyond. Heavier weight also required a higher propellant mass for a longer burn and that too added unnecessary weight. Because SpaceShipOne would be required to operate several times, and due to the encroaching interest in deploying it as a pathfinder vehicle for commercial use by fee-paying passengers, the safety of the rocket motor was paramount.

Solid propellant rocket motors are difficult to terminate part way through their thrust cycle and liquid propellant motors using hypergolic fuel and oxidiser which ignite on contact carry a greater risk of catastrophic failure, if only due to accidental mixing of the two. Liquid propellant motors using non-hypergolic propellants can be pressure-fed or require turbines to deliver the fuel and the oxidiser, complicating the design and adding hardware to the assembly. Hybrid engines are unusual because they trade outright performance in terms of thrust per unit mass of propellant (fuel and oxidiser) and use a liquid oxidiser to burn a solid-propellant fuel.

Balanced against safety and reliability, these factors narrowed the choice, along with development cost and the ease with which SpaceShipOne could be turned around for another flight during the test programme and the two-week re-flight requirement for the X Prize. There was no requirement to go beyond the 100km line and the ability of the vehicle to achieve that established the overall thrust required against the propellant mass needed for the vehicle and the motor to achieve the limited objective.

The way the rocket motor works and how it is integrated with the structural frame of the fuselage is unusual. The HTPB is contained in a narrow cylinder attached to the rear of the oxidiser tank, which forms part of the fuselage as a structural, load-bearing component. The cantilevered fuel tank allows a different size rocket motor to be attached to the end for upgrades or improvements without compromising the integrity of the fuel and oxidiser tanks. Much like the nozzle on a solid-propellant rocket motor, where a chemical mix of fuel and oxidiser is bound together, the case, the convergent-divergent throat, and the nozzle are one and referred to as the casing, throat and nozzle, or CTN.

With a composite liner, an external wrapping of graphite-epoxy material and titanium flanges, the oxidiser tank is the only part of the assembly which is structurally attached to the fuselage. Contained within a maximum width of 5ft (1.52m), it consists of a short cylinder with hemispherical end-domes, the CTN being bolted to the rear of the tank and cantilevered through the aft end of the fuselage. The fill and vent valves are located on the forward bulkhead and the tank is pressurised at $700lb/in^2$ (4,800kPa). The oxidiser tank and CTN interface is sealed with an O-ring to prevent leakage, but this is a weak part of the design, the CTN itself being a single element and thus devoid of leak points. The igniter is at the head of the CTN, and the oxidiser is pressure-fed through to the solid fuel container, avoiding the need for pumps.

Scaled Composites built the CTN liner and fuel casing with a composite wrap provided by Thiokol and an ablative exhaust nozzle from AAE Aerospace. SpaceDev provided the remaining components including the ignition equipment,

the electronic control system, the solid fuel casing, and the tank bulkheads. The solid fuel segment had four holes running the length of the container and this showed a tendency to release chunks of PBAN during combustion, a situation which would correct itself by burning the liberated material in the chamber. The solid fuel is the only part of SpaceShipOne which needed replacing on every flight, except the nitrous oxide propellant and the CTN structure.

In the original layout, the single rocket motor protruded from the back of the fuselage, but this incurred instability on the aerodynamic profile and a fairing was added which smoothed air flow across to the end of the nozzle. This incurred excess heating from the combustion chamber and nozzle which damaged the structure, prompting a change in the supporting attachment with the addition of bracing ribs and the interior face being painted white to reflect thermal flow patterns. In all cases, the nozzle, or expansion chamber, had an expansion ratio of 25:1, defined as the ratio

of the diameter of the throat to that of the outer diameter of the nozzle, a value optimised for the upper atmosphere.

The rocket motor had a maximum fixed thrust level of 20,000lb (88kN), upgraded in September 2004 from a thrust of 17,000lb (76kN) for the first several flights, the burn duration increasing from 76 seconds to 87 seconds effected through more oxidiser and an increase in the size of the oxidiser tank. Thrust decreases over the burn duration due to reducing oxidiser delivery pressure as it is consumed and through the greater proportion of the oxidiser becoming gaseous, which affects the energy output during combustion.

Control of the flight and the trajectory were managed by the pilot who received information from a System Navigation Unit (SNU) with separate display options for the four sequential phases: powered flight; coast phase; re-entry; and glide to touchdown. The SNU operated by integrating satellite GPS with sensors for position and attitude orientation, providing speed, rate of climb, rate of descent and pertinent information related to the condition of the propulsion system and to deployment of the wing/booms pivot mechanism.

Essentially an inertial navigation system, it complemented and backed up GPS-derived position and location information. Colour LCDs provided information modes which were selected by the pilot. Developed and provided by Fundamental Technology Solutions, the SNU connected SpaceShipOne with ground controllers monitoring the health of the vehicle via telemetry channels, adding a wider breadth of data than that available to the pilot.

A White Knight

The carrier-plane for SpaceShipOne, White Knight emerged from the design layout and the materials technology of the Proteus tandem-wing, high-altitude endurance aircraft designed and built by Scale Composites. Proteus made its first flight on July 28, 1998, as a high-altitude, endurance vehicle capable of remaining aloft for more than 12 hours at altitudes of up to 35,400ft (10,790m) with equipment contained within a ventral pod. With a service ceiling of 63,245ft (19,277m), it is powered by two Williams FJ44-2 turbofan engines delivering a thrust of 2,293lb (10.2kN) and carries a crew of two. Proteus has a wing span of 77.58ft

(23.65m) with an area of 300.5ft^2 (27.92m^2) and a length of 56.33ft (17.17m); the canard foreplane having an area of 178.7ft^2 (16.6m^2).

Developed and still used as a multi-purpose platform for a wide range of experiments and payloads, Proteus was the design base for the Model 318 White Knight carrier-plane and its wing owes much to that precursor. The lift requirement is set by the loaded weight of SpaceShipOne, which came in at 7,937lb (3,600kg) versus the 2,000lb (907kg) payload of Proteus. To save costs, several aspects of White Knight mirrored the outer mould line of SpaceShipOne, particularly in the forward fuselage where it could also double as a simulator for training pilots of the rocket-ship.

White Knight was required to lift SpaceShipOne to an altitude of about 49,000ft (14,935m) powered by two General Electric J85-GE-5 afterburning turbojets with a dry thrust of 2,400lb (11kN) and a wet thrust of 3,600lb (16kN). The fuselage section had a diameter of 60in (1.52m) with an interior diameter the same as that of SpaceShipOne. The nominal wing span of 82ft (25m) can be extended to 93ft (28.3m) if necessary and large speed brakes allowed

ABOVE • The Scaled Composites Proteus experimental research platform, which served to inspire the configuration of SpaceShipOne. (Scaled Composites)

ABOVE • *The White Knight carrier-plane with a forward nose section duplicating the form and interior of SpaceShipOne for doubling as a familiarisation trainer. (Don Ramey Logan)*

the same steep descent profile to those of SpaceShipOne, with a L/D ratio of almost 4.5. The pressurised cabin could accommodate up to three crewmembers in a design closely matching that of SpaceShipOne for which it doubles as a high-fidelity, moving base simulator by duplicating cockpit layout, avionics, environmental control system, trim servos, data system and electrical components.

The flight profile lifted SpaceShipOne to a nominal altitude of 8.7 miles (14km) which took approximately one hour, at which point it dropped the spaceplane which fell away for a few seconds before the pilot ignited the rocket motor for an ascent, climbing at 65° and to a maximum acceleration of 1.7g. This 'staged' sequence allowed two abort points: the first when the decision was made to release SpaceShipOne which, if a problem were indicated, could be cancelled for the joined combination to return to landing; and also after separation at the point of ignition where the pilot could elect to abort and glide down to a landing.

A normal profile would see SpaceShipOne reach an altitude slightly in excess of 100km whereupon the pilot re-configures the wings into a high drag orientation and begins the free fall back through the atmosphere. At an altitude of between 12 miles (19km) and six miles (9.6km) the pilot returns the vehicle to a low-drag configuration and glides back to the ground, followed shortly thereafter by the White Knight carrier-plane.

Preparations for powered flights began with 12 ground tests of the rocket motor, conducted between November 21, 2002 and November 18, 2003. The final ground firing was

a qualification run to clear the motor for the two Ansari X Prize flights. The airborne flight tests of the combined SpaceShipOne and White Knight vehicles would conform to a pre-planned sequence of tests and qualification runs. These were divided into captive-carry flights where the two remained joined before separation and unpowered descents to evaluate aerodynamic properties and handling characteristics prior to powered flights at short-duration motor burns working up to full-scale runs.

The first captive-carry flight took place on May 20, 2003 without a crew aboard SpaceShipOne and with the controls locked. The two pilots aboard White Knight, Scaled Composites test pilot Peter Siebold and former US Navy pilot Brian Binnie had nothing but praise for the handling qualities of the tandem configuration. The second flight on July 29 had test pilot Mike Melvill in SpaceShipOne for a flight of just over two hours, in which all its systems were turned on and evaluated while still captive to the carrier-plane.

A week later, on August 7 the third flight gave SpaceShipOne its first drop and glide test, separated from White Knight at 47,000ft (14,325m) and a speed of 121mph (194kph) with a full test of its handling characteristics from just above the stall to around 173mph (278kph) in a flight lasting 19 minutes, Melvill again at the controls. The first opportunity to translate the aft wing section and boom assemblies upwards into a feather mode came with the fourth flight on August 27 when Melvill verified that it behaved as expected. The third glide flight took place on September 23 where Melvill pushed the vehicle to both glide and feather modes and more aggressive shakedown of its performance.

A fourth glide followed on October 17 in a free fight lasting 17min 49sec where further expansion of the performance envelope was tested, and checks made of pre-ignition sequences for the rocket motor. Further evaluation on November 14 and another five days later added further tests of the propulsion system before a final drop on December 4 cleared the vehicle for the first powered flight.

That event took place on December 17, 2003, when Brian Binnie made his second flight in SpaceShipOne and lit the rocket motor at an altitude of 44,400ft (13,533m) for 18 seconds. During that period, the vehicle became supersonic for the first time, reaching Mach 1.2 and an altitude of 67,800ft (20,665m), the wing section being fully feathered and then retracted back down. On landing, 18min 10sec after release, the left main gear collapsed

BELOW • *This elegant and evocative in-flight shot of White Knight and SpaceShipOne displays the design genius of Burt Rutan and his small team of aeronautical designers and materials technologists. (Scaled Composites)*

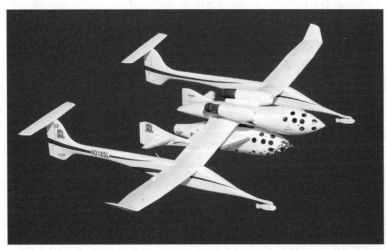

and SpaceShipOne slewed off the runway into the sand, incurring minor damage. It was back in the air for its 12th flight, an unpowered glide test lasting 18min 30sec on March 11, 2004, with test pilot Peter Siebold at the controls.

The second powered flight took place on April 8 during which Siebold fired the motor for 40 seconds to reach Mach 1.6 and an altitude of 105,000ft (32,004m), preceding a powered flight on May 13 when Melvill ignited the motor for 55 seconds, pushing through Mach 2.5 and boosting the vehicle to 150,000ft (45,720m). Some minor technical issues were noted with the avionics displays but nothing to stop the next flight on June 21 where Melvill achieved Mach 2.9 and a powered altitude of 180,000ft (54,864m) from where SpaceShipOne coasted up to a peak altitude of 328,491ft (62.2 miles/100.1km) – the first flight above the Kármán line.

The first of two flights to win the Ansari X Prize took place on September 29, several thousand onlookers anxious to see whether SpaceShipOne could start the clock to a two-week finishing line. White Knight took off with its rocket-powered spaceplane at 7.12am local time, climbing to an altitude of 46,500ft (14,173m). Flight engineer Matt Stinemetze released SpaceShipOne at 8.10am and Melvill fired up the motor, letting it run to depletion 77 seconds later to a speed of Mach 2.92, about 2,110mph (3,395kph) and an altitude of 180,000ft (54,864m) from where it continued to expend its energy, climbing to 337,700ft (102,930m).

But there had been a problem. At one minute into the burn roll rates began to increase alarmingly, reaching 190°/sec as it corkscrewed its way to peak altitude, Melvill bringing it under control first with aero-controls and then using the thrusters outside the atmosphere. Experiencing weightlessness for three minutes 30 seconds, Melvill conducted a near-perfect entry, SpaceShipOne accelerating to Mach 3 but recovering to retract the feathering at 61,000ft (18,592m) and reaching the ground 24 minutes after the drop.

To achieve the goal, win the prize and demonstrate that a commercial future awaited, the next flight had to occur before October 13 and preparations quickly prepared the two air vehicles for a repeat flight to the Kármán line and beyond. As dawn broke on October 4 a large crowd of invited guests, workers, and much of the local population gathered to watch the take-off at 6.49am, flight engineer Matt Stinemetze releasing SpaceShipOne exactly one hour later at an altitude of 47,100ft (14,356m). With Brian Binnie at the controls, the rocket-ship fired up the motor so quickly that its crackling rumble could be heard inside White Knight as it surged away on an 83 second burn accelerating to Mach 3.09, 2,186mph (3,517kph) at 213,000ft (64,922m).

Still climbing, it reached an altitude of 367,500ft (112km) where Binnie feathered the wing/boom assembly

ABOVE • Mike Melvill celebrates the first privately funded space flight after returning from above the Kármán line for the first time. (Don Ramey Logan)

LEFT • Test pilot Mike Melvill (left) chats with Scott Crossfield who, on November 20, 1953 while flying the Douglas D-588-II rocket research aircraft, became the first man to exceed Mach 2. (Don Ramey Logan)

BELOW • A fine shot of the main landing gear and doors with a twin-wheeled nose dolly below the skid for moving SpaceShipOne around. (Don Ramey Logan)

before commencing the descent back down through the atmosphere, accelerating to Mach 3.25 and a peak deceleration of 5.4g at 105,000ft (32,000m). The Ansari X Prize had been won in a free flight lasting 24 minutes, out-climbing the highest altitude reached by the storied X-15, a height of 354,200ft (107,960m) on August 22, 1963 in a flight piloted by Joe Walker.

In support of SpaceShipOne, White Knight had made 17 flights between May 20, 2003, and October 4, 2004, launching six powered runs for the rocket-ship of which three were beyond the Kármán line. News spread far and wide and Virgin Galactic made much of the spectacular achievement through a simple, elegant path to eclipse the expectations of many. Already ahead of the game, Richard Branson was outpacing the negative opinions of naysayers. But it would take more than confidence to transform a technical achievement into a commercial success.

FLOATING FREE

Winged spaceplanes for hire

RIGHT • WhiteKnight was used to carry early models of the X-37, formerly NASA's Orbital Test Vehicle as it transitioned from the space agency to the US Air Force. (NASA)

After its role in winning the Ansari X Prize, Scaled Composites was contracted by the US Defense Advanced Projects Agency (DARPA) to use White Knight for captive-carry and test flights with the Boeing X-37A. The first was on June 21, 2005, used to lift the spaceplane into the air for evaluation without release for a free flight. The first glide flight took place on April 7, 2006 when it was released from White Knight in a successful test of its aerodynamic properties, but an accident during landing caused minor damage. After operations moved from Mojave to Plant 42 in Palmdale, California, two additional drop-tests were completed on August 18 and September 26, 2006.

Although unmanned, the X-37 originated as a NASA programme in the late 1990s to provide a cheaper means of lifting relatively small payloads into space without using the costly Shuttle, the initial application being to lift it into space inside the payload bay and place it in orbit. With aerodynamics dictated by the Shuttle itself, it was originally known as the Orbital Test Vehicle with many potential applications including being able to operate in space and repair satellites using automated equipment before returning to Earth. That application was rendered redundant when NASA decided to retire the Shuttle, but the US Air Force took it up and the original vehicle was modified into the X-37B which is launched on highly secret missions by a conventional expendable launch vehicle.

BELOW • Redundant for its primary task of carrying SpaceShipOne, the WhiteKnight carrier-plane was hired out to a range of customers for experimental duties, seen here carrying a Northrop Grumman radar pod. (Alan Roderick Akradecki)

White Knight played a highly significant part in this programme, which had finally given the air force a reusable spaceplane capable of a wide range of science and technology tests and applications, all of which bar a few benign experiments are classified. The first X-37B was sent into space by an Atlas V rocket from Cape Canaveral on April 22, 2010 and remained in orbit for almost 225 days before returning to a successful landing. Since then, the two spaceplanes have conducted seven long-duration flights, some of them on the commercial Falcon 9 rocket operated by SpaceX, the most recent launched by Falcon Heavy on December 29, 2023 and which continues on its mission. Don't hold your breath waiting for its return; the longest mission to date was the sixth, launched on November 12, 2020, lasting 908 days.

White Knight played a similarly pioneering role in the US Air Force's Mission Adaptive Compliant Wing (MACW) programme which sought to test a trailing-edge flap that could change camber according to various flight conditions to increase effectiveness and improve performance with existing aircraft and future designs. Flights occurred between October and December 2006 where the carrier operated as a research platform into a wide range of different atmospheric conditions, providing valuable data on how in-flight changes to wing shape could make aircraft more efficient and improve performance. But it is with space-related commercial programmes that it is best remembered.

The ability to build comparatively small spaceplanes as demonstrated by the government-funded X-37A and X-37B programmes inspired entrepreneurs to apply the lessons learned and use the aerodynamic research into winged, recoverable space vehicles to design commercial projects. The most visible today is the Dream Chaser, now close to flying on its first test in space, about which more later on. In pioneering the use of such designs and as a carrier for SpaceShipOne and the X-37, White Knight has a solid place in the pantheon of pioneer aviation. Recognising its value for subsequent generations to appreciate, it was retired to the Paul Allen Flying Heritage Collection in Everett, Washington, on July 22, 2014.

A New Generation

Burt Rutan and Richard Branson unveiled plans for the commercial, passenger-carrying SpaceShipTwo (SS2) and its carrier-plane White Knight Two (WK2) at the American Museum of Natural History, New York, on January 23, 2008. President of Virgin Galactic, Will Whitesides said that this was "part of a much longer development programme" which would give the SS2 an open architecture like Linux to allow other people to develop new vehicles and revolutionise "new industrial uses of space." Released in September 1991, Linux is a computer operating system which became the dominant application for Android smartphones. Clarifying, Whitesides claimed that "we will work with people...to do new things such as White Knight Two's wing to build new aircraft".

Outlining future plans, Virgin Galactic claimed that the programme anticipated building up to 50 rocket-ships, supporting flights twice a day with a safety standard that Whitesides defined as "100 times more safe than government space travel." It was, said Rutan, "not a small programme," raising expectations which were levelled by the pragmatic Burt Rutan in reminding his audience: "Don't believe anyone that tells you that the safety will be the same as a modern airliner, which has been around for 70 years."

At the rollout of plans which was also attended by several future travellers, it was expected that flight tests would start by 2008, with operational space flights from New Mexico's Spaceport America the following year, replacing the Mojave Air and Space Port for Virgin Galactic operations that had been the mainstay of operations with SpaceShipOne. Technical challenges and fatal accidents would take their toll on decisions and on programme confidence as changes in schedules extended the development timeline.

While the overall flight profile was similar to that flown by SpaceShipOne and WhiteKnight, the design of the two successors was completely different, dictated by the philosophy of light weight and simplicity and by the need to carry six passengers in addition to the two pilots, both of whom were required to share the workload and for the co-pilot to attend to any of the passengers requiring assistance. The reaction of individuals to the experience of flight to the edge of space was unpredictable, none of whom would have had the same familiarisation flying in aircraft of this type or experiencing rocket-powered vehicles and their relatively high g-load.

Moreover, the psychological reaction to this experience was equally unpredictable, potentially ranging from high levels of euphoria to uncontrollable panic. Some studies had been conducted on passenger reactions in conventional airliners and it was believed that 10% of those booked aboard could experience reactions they had never encountered before. Hence the need for the co-pilot to handle such situations.

As Model 339, the SS2 rocket-ship was designed as a low-wing, twin-tailboom vehicle with an outboard horizontal tail like its predecessor. The tricycle landing gear has gravity-drop deployment and no retraction but with a single nose wheel rather than the skid on SpaceShipOne. Composite materials are used extensively throughout the primary structure of the nose, the pressurised cabin, aft fuselage, wings, feather flap assembly and the horizontal stabilisers. To herald the dawn of a new era in rocket-ships from Scaled Composites, the first SS2 was named VSS (Virgin Space Ship) *Enterprise* and registered N339SS.

Overall, SS2 has a length of 60ft (18.28m), a wing span of 27.25ft (8.3m), an overall height of 18.08ft (5.5m) and a gross weight of 21,473lb (6,545kg). With a diameter of 7.5ft (2.2m), the fuselage is about 50% wider than that of SpaceShipOne with the two pilots seated side-by-side and the six passengers seated behind, three each side. In reality, no more than four

ABOVE • As envisaged by Burt Rutan and Richard Branson, SpaceShipTwo and White Knight Two were a considerable investment in a passenger-carrying, suborbital capability. (Virgin Galactic)

LEFT • The uncrewed X-37 evolved into the X-37B sent into space by expendable launch vehicles on long-duration missions carrying a wide range of experiments. (USAF)

*BELOW •
SpaceShipTwo
incorporated
a redesigned
feather system
for deceleration
on re-entering
the atmosphere.
(Virgin Galactic)*

passengers were ever carried. The pressurised forward cabin has a length of 12ft (3.7m). Two sets of windows are provided, six at 17in (43cm) diameter with three each side of the fuselage and three with a 13in (33cm) diameter along the top.

Primary aerodynamic controls consist of elevons for pitch and roll and rudders for yaw, electrical power being provided by two primary battery packs and one emergency pack. Four high-pressure bottles, two in each wing leading edge, supply compressed air to a range of systems including the feather mechanism. Thrusters are employed for attitude control outside the atmosphere, but Scaled Composites claimed that the vehicle could re-enter at any orientation and stabilise itself.

The feather system for high drag adopted for SpaceShipOne is a little different for SS2 in that it the later design incorporates the aft one-third of the wing surface and the tailbooms, pivoting upwards to an inclination of 60° to slow the vehicle during re-entry. It is attached to the wing rear spar at two outboard and two inboard hinges, located on the wingtips and the aft fuselage, respectively. They are locked into the normal down position by hardened steel pins with left and right actuators mounted in the aft fuselage, with the deploy handle inside the crew cabin.

To avoid the catastrophe of a premature deployment, the handle was force-loaded so that there could be no premature release of the mechanism. The optimum speed for deploying the feather systems was between Mach 1.4 and Mach 1.8 so that the crew could test the system to see that it worked while allowing time to shut down the rocket motor and abort if it was jammed or inoperable; re-entry without the feather system would be catastrophic.

The rocket motor for the SS2 ships, known as RocketMotorTwo (RM2), retained the same hybrid operating concept using HTPB and nitrous oxide propellants but in May 2014 Virgin Galactic took the development of the more powerful motor in-house, terminating its contract with Sierra Nevada Corporation. The fuel was changed from HTPB to a thermoplastic polyamide formulation which was expected to provide better performance, slightly increasing the thrust and the projected altitude reached by the SS2 rocket-ship. RM2 has a thrust of 61,000lb (271.32kN) and a burn duration of 60 seconds.

Hot fire tests had been conducted from June 2005 but the danger in dealing with volatile propellants was made all too obvious on July 26, 2007, when a standard test was

was developed by Scaled Composites between 2007 and 2010, three times the size of its precursor and with four engines instead of two. Taking a leaf out of the Proteus applications book, it too was conceived as a multi-purpose cargo-lifter which could also be applied to research tasks for a wide range of customers. For that it had an open-architecture.

Under the original business plan, there were to be two WK2 airframes and five SS2 rocket ships, only the initial example of each type being assembled by Scaled Composites, the others by Virgin Galactic as it sought to take ownership of manufacturing and operation. The design incorporated commercial marketing expectations, with two fuselage pods connected by a centre wing structure and powered by four jet engines, two under each wing on dedicated pylons. It would have a service ceiling of 60,000ft (18,288m).

One fuselage was for piloting the aircraft with an interior and a profile identical to that of SpaceShipTwo for familiarisation flights, the other fuselage capable of carrying passengers to the edge of the stratosphere where the curvature of the Earth can be seen and the sky is blue-black, an experience those flying in Concorde had before it was retired in October 2003. Virgin Galactic planned to offer 'cut-price' tickets for the 'Concorde-experience' but that economy-class option never materialised.

The design of White Knight Two was revealed on January 23, 2008 and the first of a planned two aircraft was rolled out to great acclaim and celebration on July 28, bearing the name VMS (Virgin Mother Ship) *Eve*, after Richard Branson's mother Evette who was also chairperson of the

LEFT • SpaceShipTwo (bottom) was twice as big as its predecessor, SpaceShipOne, with a capacity for up to six passengers, although in operation it would never carry more than four. (Virgin Galactic)

BOTTOM • Attachment of the forward fuselage section to the mid-fuselage, completely revised from the SpaceShipOne configuration. (Virgin Galactic)

BELOW • Construction of the fuselage sections for the Enterprise *shows the mating of upper and lower centre fuselage sections and the orientation with the aft section. (Virgin Galactic)*

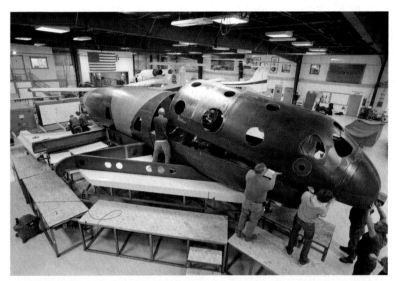

conducted by Scaled Composites at the Mojave Air and Space Port. After filling the oxidiser tank with 9,900lb (4,490kg) of nitrous oxide for a planned 15 second cold-flow test of the injector, contamination caused the tank to explode, killing three engineers and injuring three more as shrapnel showered the area. Hot fire tests of RM2 began on April 20, 2009, with the first full-scale firing on June 20, 2012. Scaled Composites conducted a test on May 17, 2013, with flaws intentionally introduced to see if the motor would remain stable, contributing data to the overall safety margins.

But development slowed to a crawl and while captive-carry and glide flights got under way with VSS *Enterprise*, only in December 2012 did Virgin Galactic conduct a glide test with the rocket motor installed. After several test runs and a modest redesign of the fuselage to incorporate additional tanks of methane and helium in the wings, fresh ground runs were completed by October 2014, but problems persisted and late in 2015 Virgin Galactic took the decision to revert to the HTPB fuel following a long and exhaustive examination of all potential failure modes and risk factors.

This review of all elements of the programme came in the months after the loss of the first SS2 rocket-ship on October 31, 2014, discussed in detail later in this section. There were marginal risks associated with the new propulsion unit, although the rocket motor was not implicated in the accident.

A New White Knight

To put SS2 at the correct altitude for its rocket-run to the Kármán line and beyond, a carrier-plane would be required with greater carrying capacity and performance than WhiteKnight possessed. For that, White Knight Two (WK2)

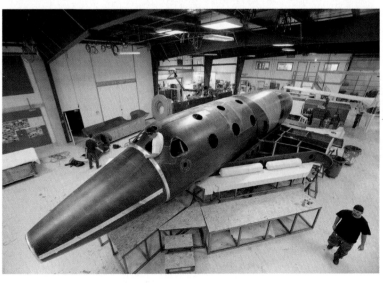

RIGHT • 'Enterprise' rolls out, displaying its size and overall design configuration with the tail booms in the glide position. (Virgin Galactic)

BELOW • The tail section of 'Enterprise' displaying the arrangement of the two rudders and horizontal control surfaces. (Virgin Galactic)

Virgin Group. Bearing the registration number N348MS, it was impressive in size and potential. *Eve* had a length of 78.75ft (24m), a wingspan of 141.08ft (43m), powered by four P&W Canada PW308 turbofan engines with a thrust of 6,900lb (30.69kN). With a rated payload capacity of 37,479lb (17,000kg), in addition to carrying SS2 rocket-ships, it was designed to also carry a single expendable rocket for Branson's Launcher One project sending satellites into space, but it would never actually do that.

The first flight took place on December 21, 2008, which lasted one hour followed by a series of tests which explored the flight envelope. On June 20, 2009, it performed flybys for spectators during the ground-breaking event for Spaceport America in New Mexico. During the remainder of the year, VMS *Eve* supported several air shows and publicity events across the United States, including Oshkosh and open days at Edwards Air Force Base, important and highly visible appearances to enhance awareness of progress and recruit potential flight applicants, not all of whom would translate through to firm orders due to medical reasons and other factors.

The first captive-carry flight with *Enterprise* occurred on March 22, 2010, the 'mated' configuration remaining aloft for almost three hours. The next flight on May 1 evaluated a new speed brake and on the third captive-carry flight on May 16 various systems on *Enterprise* were tested including pressurisation and electrical equipment along with the avionics.

On July 15, test pilots Siebold and Alsbury were on board *Enterprise* for the first time in a flight lasting six hours 21min to check out systems, with *Eve* flying a simulated post-release flight pattern.

This cycle of simulated descent and landing profiles continued over the next several flights in the mated configuration but without a crew in *Enterprise* until the 39th flight of *Eve* on September 30, 2010, when Siebold and Alsbury were again on board for a final rehearsal lasting five hours before the first drop and glide flight, which would occur two flights later.

While *Enterprise* and rocket motor development progressed, Virgin Galactic's CEO Will Whitehorn presided over setting up the detailed flight test programme, which was to be greatly expanded over that conducted for the Ansari X Prize, elements of which had only been necessary to receive an FAA licence to fly experimental runs. This was to be very different. A licence to operate a commercial service and fly paying passengers required a broader range of checkpoints, activity akin to operating an airline. The flight test phase for *Enterprise* would call for exceptional rigour in verifying to the satisfaction of the regulatory authorities that the vehicles were safe and risk free, to a very high level.

The test programme was divided into seven separate stages: ground tests; captive-carry flight; glide drops; subsonic powered flight; supersonic tests in the atmosphere; suborbital

space flight; and consultation with the FAA to obtain a commercial licence. The licence would also certify the launch site for take-off and landing, the facilities supporting flight preparation and operation and certain requirements for safety factors and checks. Government spacecraft and launch vehicles do not require FAA licences, since they are subject to separate NASA and US Air Force regulations but any commercial flight, be it for a space vehicle or a launch system, is required to be inspected and certified by the civilian FAA.

Enterprise Flies

The first glide flight for *Enterprise* was achieved on October 10, 2010, crewed by Siebold and Alsbury in a flight lasting 13 minutes from dropping out of the restraining shoe underneath *Eve* at an altitude of 46,000ft (14,020m) to landing back at the Mojave Air and Space Port. Between this date and June 27, 2011, a total of 15 glide flights were completed but on June 9 one drop test had been aborted when the release mechanism on *Eve* suffered a technical failure. Glide flights had provided familiarisation for other pilots, including Doug Shane, Brian Binnie, who had flown SpaceShipOne on the second and qualifying flight for the Ansari X Prize in October 2003, Mark Stucky and Clint Nichols.

Much had been learned from the glide flights and extensive testing with both *Eve* and *Enterprise*, but the engineers wanted time to analyse the data and to make a number of changes to equipment and to tweak flight procedures, refining pilot's notes, and handling instructions as well as engineering procedures for turning around the vehicles between flights. The programme had already demonstrated two successful glide flights within a 24-hr period, but for three months the fleet remained on the ground in preparation for a resumption of tests, progressing toward powered flight.

On September 29, 2011, the programme resumed with a heart-stopping moment when *Enterprise* plunged unexpectedly into a steep descent after encountering a stall condition on the tail. Corrected by the feather system, when the descent rate had been arrested, it was returned to glide

mode but a free descent that should have taken around eleven minutes to land, put *Enterprise* on the ground in seven minutes 15sec. It was a cathartic moment for former NASA executive Mike Moses who had just joined the team as vice president of operations, focusing especially on operational safety and risk management.

Test flights resumed on June 26, 2012 with a flight lasting 11 minutes 22sec and incorporating some of the plumbing for the Sierra Nevada hybrid engine. With a claimed list of 515 people now registered to fly as passengers and having paid deposits, expectations were high that the first fee-payers could experience weightlessness the following year. Over the next two months five additional glide flights cleared test objectives related to airspeed, centre-of-gravity, and structural loads analysis. The first flight with the new rocket motor and a shorter aft skirt was completed on December 19 and a further two glide drops set up *Enterprise* for the next phase.

The first powered flight of VSS *Enterprise*, began at 7.02am on April 29, 2013, when Mark Stucky and Mike Alsbury took off from the Space Port and separated from VMS *Eve* 45 minutes later at an altitude of 47,000ft (14,325m). The burn lasted 16 seconds, propelling SS2 to a

ABOVE • A rocket motor test run conducted at the Northrop Grumman facility in 2009. (Northrop Grumman)

BELOW • A view of the aft quarter-section of Enterprise *on a glide flight with a rocket motor tail cone and aerodynamic cover. (Virgin Galactic)*

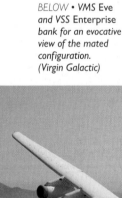

height of 55,000ft (16,764m) and a speed of Mach 1.2, Stucky piloting the vehicle to a good landing about 10 minutes after separation. The next flight on September 5 achieved Mach 1.43 and reached 69,000ft (21,031m) followed by the third powered flight on January 10, 2014 piloted by Virgin Galactic's chief test pilot Dave Mackay who pushed the altitude to 71,000ft (21,640m). Confidence was high with expectations that *Enterprise* would reach space that year. The next flight would change that.

After a nine-month hiatus, SS2 made its fourth powered flight on October 31, 2014, equipped with the new rocket motor from Sierra Nevada, that configuration preceded on October 7 by an unpowered glide test to check on aerodynamic handling. The pilot for this inaugural test of the new motor, Pete Siebold was highly qualified and certified as the pilot-in-command for *Enterprise*, *Eve*, Proteus, and multi-engine aircraft. Siebold had worked for Scaled Composites since 1996 as a design engineer, joining the SS2 programme in 2010 to also become its director of flight operations, logging 49 hours in the rocket-ship. The co-pilot, Michael Alsbury was an airline transport pilot hired by Scaled Composites in January 2000 as an engineer, transitioning to the SS2 programme in which he had completed seven glide flights.

Events for the fourth powered flight began when the flight crews for both air vehicles arrived for a briefing at 5am on October 31, 2014, and after a short break they walked across to the staging area for pre-flight preparations at 7.30am. The *Enterprise* flight crew got inside their rocket-ship at 8.15am and the mated configuration became airborne at 9.19am, the pre-planned series of checks prior to the drop beginning at 9.58am. A second checklist at L-4min began at

10.03am, followed two minutes later by an exchange between *Enterprise* and *Eve* on precise sequences for the drop and ignition of the rocket motor.

A series of checks verified the correct position for the SS2's control stick, moved forward for a nose pitch-down to separate from the carrier-plane, with verification that the launch pylon switch was armed, its release button in the cockpit glowing orange. The final verbal countdown to release began at 10.17.16am with separation actuated from *Eve*'s flight deck three seconds later and with the words 'clean release'.

At this point *Eve* was at an altitude of 46,400ft (14,142m) and climbing to maximise vertical separation before levelling off. Siebold ordered Alsbury to fire the rocket motor which ignited at 10.07.21am followed five seconds later by the co-pilot calling out that the speed had reached Mach 0.8, routinely alerting the pilot that at this point there would be a transonic pitch-up as the centre of lift shifted forward, followed by a pitch down as it shifted back. All normal and predictable.

Right at this point, under standard procedures the pilot would trim the horizontal stabilisers located on the outboard side of each tail boom to -14° pitch and the co-pilot would then unlock the feather lock handle when the ship reached Mach 1.4. There was a finite window from Mach 1.4 to Mach 1.8 during which the feather lock had to be released. Later than that and the flight would have to be abandoned.

The on-board image recorder showed the handle actuated at Mach 0.82, 9.8sec after release from *Eve* and at least 13 seconds before the rocket-ship would have reached Mach 1.4, the earliest point for deployment. As post-flight analysis of the data would show, premature release of the lock handle caused the feather system to begin moving at 10.07.30am, about 11 seconds

after release from the carrier-plane, and that caused *Enterprise* to break up two seconds later at an altitude of 46,000ft (14,020m).

During the break-up, with the sound of the structure physically tearing apart and still strapped to his seat, dressed only in flight coveralls, Siebold was thrown from the rocket-ship into clear air at a speed of about 540mph (869kph). Momentarily losing consciousness in the thin air, he came to with the vast expanse of the desert floor spread out far below. After unbuckling his straps to allow the seat to fall away, Siebold again lapsed into unconsciousness before his personal parachute opened at 11,590ft (3,532m), lowering him slowly to the ground. Alsbury was unable to release himself from *Enterprise* and perished when the wreckage fell to the desert.

A Tsunami of Opportunities

Before the accident to *Enterprise*, Branson had expected commercial flights to begin within 18 months, but a range of factors conspired to shed a cloud over that plan. Construction of the second SpaceShipTwo vehicle had begun in 2012 and in September 2014 it had been filed with the FAA and named *Unity* carrying the registration N202VG. Assembly went quickly after the loss of *Enterprise*, a universal resolve to press ahead, begin flight tests and get fee-paying passengers lifted from the wait list. At the time of the accident *Unity* was 90% complete structurally and 65% complete as a working air vehicle.

Focusing on what they considered to be a niche service for space tourists, on February 19, 2016, *Unity* was unveiled with ground tests commencing immediately. Virgin Galactic and The Spaceship Company planned a reduced flight test phase compared to that conducted with *Enterprise*. Much of the basic concept demonstration and evaluation

had already been performed on the first SS2 vehicle and less than half the number of flights were now required in each test regime for *Unity* to reach qualification for powered flight.

A series of four captive-carry flights were conducted between September 8 and November 30, 2016, in which period two planned glide flights had to be cancelled due to high winds. Piloted by Mark Stucky and David Mackay, the first drop test took place on December 3, a flight lasting 10 minutes and reaching an altitude of 55,000ft (16,764m) and a speed of Mach 0.6. In association with the second glide flight on December 22, Virgin Galactic announced that 700 customers had signed up for a flight. The last of seven glide flights took place on January 11, 2018.

By this date, Branson still nursed hope of developing the fleet into point-to-point vehicles where flights take place directly between two destinations without going through a central hub. But hopes of a Middle East launch capability, nurtured by its principal investor Aabar, were disappearing

BELOW • In a hangar named FAITH, on February 19, 2016, VSS Unity *was rolled out to begin flight tests and obtain a licence to carry fee-paying passengers. (Virgin Galactic)*

ABOVE LEFT • With the feathered aft wing section and tail booms deployed to the re-entry position, Unity drifts to apogee on its first commercial flight. (Virgin Galactic)

ABOVE RIGHT • Unity's rocket motor roars into life on June 29, 2023, the first commercial flight carrying four members of the Italian Air Force. (Virgin Galactic)

BELOW • Displaying the Italian flag, the four passengers on Unity's inaugural commercial service enjoy four minutes of weightlessness. (Virgin Galactic)

fast. Hopes too were broaching reality with plans to develop an air-launched satellite capability by converting a redundant airliner and carrying a solid propellant rocket and its payload under its inboard wing section. That story is told in the following chapter.

Back at Mojave, plans for the first powered flight of *Unity* had been cleared by the preceding glide tests and that took place on April 5, 2018 when Stucky and Mackay piloted the rocket-ship to an altitude of 84,300ft (25,694m) and a speed of Mach 1.87. A second powered run followed on May 29 when the two pilots reached 114,501ft (34,800m) and a speed of Mach 1.9, the fuselage now carrying the six seats it would have for passenger rides. The third flight, on July 26, reached 170,800ft (52,059m) and Mach 2.47 with Mackay accompanied by co-pilot Michael Masucci. This was followed on December 13 by a flight to 271,392ft (82,720m) and Mach 2.9 piloted by Stucky and Frederick W Sturcow, a former NASA astronaut with four Shuttle missions in his flight log and who had been recruited by Virgin Galactic in 2013.

Unity carried a third crewmember on the next flight launched on February 22, 2019, when Mackay and Masucci were accompanied by Beth Moses, described by the company as 'chief space flight participant instructor and interiors program manager'. Beth would make six flights in *Unity*, three of which were on fee-paying commercial flights to assist with the passengers, a role which had been assigned to the co-pilot on *Enterprise*. Launched at 8.57am, this fifth powered flight reached 295,257ft (89,99m) and Mach 3.04 and lasted 26 minutes.

The next flight on May 1, 2020, was the first from Spaceport America, Virgin Galactic's new operating headquarters in southern New Mexico developed through a partnership with the State legislature and with the Rocket Ship Company as the anchor tenant. This was a glide flight

to qualify the facility and a second on June 25 cleared it for the first powered flight attempted from the new location.

That attempt began with take-off at 8.24am on December 12, 2020, carrying three science payloads for NASA. Unlike passenger flights where *Unity* would be inverted to the horizon so that occupants could get a view of the Earth through the upper windows, on this flight the rocket-ship would pitch 270° into re-entry attitude. The drop came at 9.15am and the motor was ignited as planned but a computer malfunction caused shutdown one second later, *Unity* being recovered to a safe descent and landing. The next attempt on May 22 was a successful powered flight to 292,776ft (89,238m) and a speed of Mach 3.

What happened next was destined for controversy. Dubbed the 'Billionaires race for Space', its crew complement was a reaction to claims by Jeff Bezos on June 21, 2021, that he would be the first founder of a commercial spaceflight programme to ride into space, using the New Shepard rocket developed through his company, Blue Origin. Not wanting to be outdone, Branson immediately filed a FAA request for *Unity* to be licensed for passenger flights, which would be piloted by Mackay and Masucci, other occupants being Sirisha Bandia, Colin Bennett, Beth Moses, and Branson himself.

For this historic flight, VMS *Eve* took off at 8.40am local time on July 11, 2021, carrying *Unity* to its drop height where it was released at 10.04am for a powered ascent to a maximum altitude of 282,277ft (86,038m), two minutes 38sec after it was released. The occupants experienced about four minutes of weightlessness and *Unity* was back on the ground after a free flight lasting 14 minutes 17sec. By exceeding a height of 50 miles (80km), it had qualified the occupants for US astronaut wings, far below the 100km Kármán line for an internationally recognised space flight. However, during the flight, *Unity* had deviated from its scheduled ground track which placed it outside the glide cone on re-entry, triggering concern at the FAA.

News that former NASA astronaut and test pilot Mark Stucky was sacked by Virgin Galactic on July 19 after publication of his book on May 4, which was heavily critical of the company's safety culture, attracted public attention. Stucky was also critical of the way the deviation on that recent flight had not been corrected by Mackay, adding further comment on social media channels. Responding, the FAA expanded the allowable glide cone for future flights to allow some acceptable deviations, which had never been a threat to other aircraft although it had strayed outside the licensed zone notified to other operators in the region. But the additional safety requirements imposed by the FAA took far longer to implement.

In qualification of those requirements and as a final set of performance data-gathering objectives prior to full commercialisation, a glide flight from Spaceport America was conducted on April 26, 2023. Released for free flight

at 47,000ft (14,325m), *Unity* took nine minutes to get down to the ground. Piloted by Masucci and Sturckow, the final flight before full release for commercial operations took place on May 25, almost two years after the previous flight to space, where it achieved an altitude of 286,176ft (87,226m). The culmination of 17 years of work and almost $1bn of investment, the company had more than $900m in cash and securities, the future was assured.

By now the company had 600 reservations for flights but the price had gone up. In 2021 it reopened sales from when seats were priced at $400,000, while the achievable altitude had gone down to well below the 100km which qualifies for international recognition as a space flight. In the six commercial flights that followed between June 29, 2023 and January 26, 2024, all were at or around the altitude reached by the final qualification flight and below the Kármán line. Each flight carries only four passengers, three if you subtract the 'participant instructor', and two pilots.

Flying with Virgin Galactic may not be as gruelling as that experienced by professional astronauts in government programmes, but the preparation is mapped in detail and requires several days of pre-flight preparation. Applicants are required to complete a detailed medical questionnaire and then to attend a series of presentations and briefings. These take place at Spaceport America and applicants meet with fellow passengers on an assigned flight and include descriptions of the two vehicles, the flight plan, safety procedures, what to expect on a normal flight, what to do in the event of an emergency and how to unbuckle and enjoy weightlessness at the apogee of the flight.

In the flight readiness programme, applicants have an additional opportunity to experience zero-g by flying a hyperbolic trajectory in an aircraft to gain awareness of the sensations and how not to be alarmed at the physical reactions. But that is an optional addition to the programme, which does include a guided tour of the facilities where the two vehicles are prepared for flight. All those who fly are accepted into the astronaut ranks of the Association of Space Explorers, which was formed in 1985 to include those who had completed one orbit of the Earth, but which will now acknowledge the Virgin Galactic suborbital flights. Currently, the registry contains 615 names of those who have conducted flights into space since Yuri Gagarin on

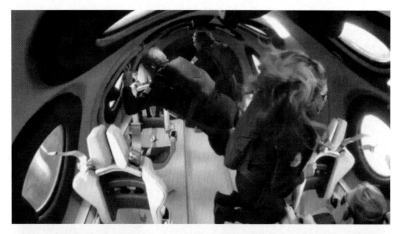

April 12, 1961. Reunions are held annually, this year's event taking place in Clear Lake Texas in December.

Beyond *Unity*, Virgin Galactic plans to fly perhaps two more commercial flights in 2024 before retiring it and shifting resources into the development of the Delta-class vehicle which it hopes to have flying in 2026. Late in 2023, Virgin Galactic laid off 18% of its workforce and focused on a new generation which had formerly been identified as SpaceShipThree, the first of which (VSS *Imagine*) was unveiled on March 30, 2021. Repurposed into Delta, the new vehicle will have capacity for six passengers, a faster turnaround and achieve higher altitudes.

ABOVE • Alan Stern, manager of NASA's New Horizons mission to the planet Pluto, rides the fifth commercial flight on November 2, 2023, along with astronaut instructor Colin Bennett, Kellie Gerardi and Ketty Maisonrouge. (Virgin Galactic)

TOP • The fourth commercial flight sent aloft on October 6, 2023, gives passengers Beth Moses, Ron Rosaro, Trevor Beattie, and Namira Salim the ride of their lives. (Virgin Galactic)

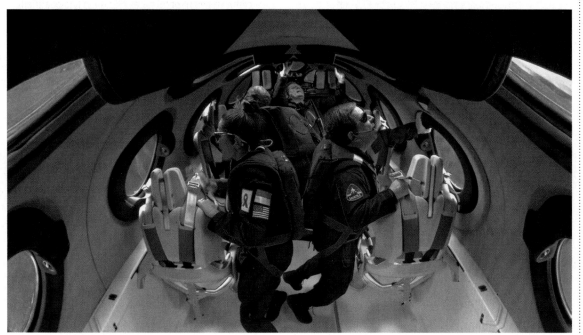

LEFT • Launched on January 26, 2024, the sixth commercial flight sends Lina Borozdina, Robie Vaughan, Franz Haider and Neil Kornswiet to the edge of space. (Virgin Galactic)

LAUNCHER
ONE GOES NOWHERE

A ride share to space collapses

The road to commercial space activity paved by Richard Branson had been long, painful, costly, and compromised by four fatalities. The costs in all areas had been far beyond what had been projected when he first made the commitment to build a passenger-carrying rocketship with wings. But it had been achieved, and the success cannot be measured in revenue alone, for having developed at tremendous expense what people around the world have dreamed of for centuries, he can offer at least a few people that ultimate high flight to the edge of space.

Commercial space flight activities are driven by market need and by the potential for innovative solutions to niche opportunities. One of those was the business of launching very small satellites at low cost for a wide range of users lining up to get their payloads into space and seeking the cheapest price on offer. Because rockets are complex, they are expensive to build, which drives price based on unit costs.

By sending up very small satellites, more organisations can get experiments or revenue-earning payloads into space which expands development and increases demand. As space technology advanced it pushed miniaturisation in both satellites and their payloads; the satellite 'bus' providing power, control systems and communications became cheaper and lighter while the 'payload', consisting of experiments or revenue-earning services, became smaller and cost less to develop. Overriding everything was the cost of launching the satellite; the larger and heavier it is, the more it costs to send it into orbit.

Rocket scientists know that launching from the ground up through the atmosphere has great disadvantage over launching from an aircraft at high altitude. This is because the density of the atmosphere at 39,000ft (11,887m) is only 18% that at the surface. The significantly reduced drag on a rocket launched from that altitude requires less thrust to achieve orbit with an engine that gains added performance from being launched at that altitude. This is because the shape of the rocket's nozzle is critical for optimum performance. In a ground-launched rocket the shape of the nozzle is a compromise between operating under high pressure at launch and much lower ambient pressure when its work is done.

There is very little reduction in the force of gravity (which declines on the inverse square law) by launching at altitude, but the combined advantages of reduced drag and better nozzle shape adds measurably to the performance of even a small rocket. Added to which is a further advantage in the speed of the aircraft at the time of launch and the greater efficiency in the use of air-breathing engines on the carrier-plane to do the work that the rocket's motor would otherwise have to do to achieve the same speed and altitude.

In the early days of the space programme, expendable rockets such as NASA's solid-propellant Scout launched small satellites from the early 1960s until its retirement in 1994. But the logic of launching satellites into space from aircraft got a public airing in 1984 when, in response to Soviet anti-satellite weapons, the US Air Force fired a missile from an F-15 Eagle knocking out a redundant American satellite in a test of its own, air-launched system.

Early in 1987, Orbital Sciences Corporation began commercial development of a satellite launcher carried

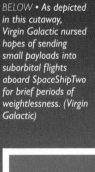

RIGHT • A Pegasus rocket drops away from the converted Lockheed L-1011 TriStar and fires its solid propellant motor on its way toward orbit. (NASA)

BELOW • As depicted in this cutaway, Virgin Galactic nursed hopes of sending small payloads into suborbital flights aboard SpaceShipTwo for brief periods of weightlessness. (Virgin Galactic)

aloft by adapted aircraft with the aim of setting services to the expanding market. A team led by Antonio Elias developed the three-stage, solid propellant Pegasus rocket which would initially be carried by a NASA B-52 which it owned for operation as a multi-purpose carrier-plane. Later, Pegasus would be carried by a converted Lockheed L-1011 TriStar airliner, renamed Stargazer, to a height of about 39,000ft (11,887m) where it was released for flight.

Weighing 40,790lb (18,500kg), and with wings for stability, Pegasus would freefall for five seconds before pitching nose up, igniting the first stage rocket motor delivering a thrust of 110,000lb (489kN) for 69 seconds, reaching a height of 200,000ft (60,960m) where the second stage ignited producing a thrust of 25,800 (114.7kN) for 75 seconds. The third stage fired for 68 seconds delivering a thrust of 7,200lb (32kN) to place a payload of up to 976lb (443kg) in orbit. Slightly more powerful variants were developed with an optional fourth stage.

The first commercial flight occurred on April 5, 1990, when the NASA B-52 was used to launch a scientific payload and by April 2021 it had flown 45 missions carrying almost 100 satellites into orbit. With a launch cost of $40m, Pegasus offered a price of $40,980/lb ($90,360/kg) of payload. Ever the consummate businessman, Richard Branson saw an opportunity to capitalise on a market niche for delivering low-cost satellites into orbit and doing that at a knock-down price.

Branson's Gamble

A late convert to space commercialisation, Branson began work on a competitive air-launched delivery system in 2007 with a plan to use the White Knight Two carrier-plane developed for SpaceShipTwo for lifting a two-stage, solid propellant rocket in a similar concept to the Pegasus programme. Marketing for what was called LauncherOne targeted both satellite manufacturers and operators with feedback showing considerable interest based on a competitively priced system. Satellite bus manufacturers such as the UK's Surrey Satellite Technology and Sierra Nevada Space Systems in the US

expressed interest in building platforms which could host payloads and packages specifically for LauncherOne.

In 2011, a new start-up operation spiked those plans when Dynetics examined the possibility of building a very large aircraft similar to White Knight Two for air-launching Falcon 9 rockets from SpaceX, formally announcing its commitment that December. Scaled Composites did the design and manufacturing of the Model 351 Stratolaunch carrier-plane which, at 385ft (117m), had the longest wingspan of any aircraft ever built. This prompted changes to the LauncherOne operation.

On July 11, 2012, at the UK's Farnborough Air Show, Virgin relaunched its small satellite business quoting a cost of no more than $10m to launch a 441lb (220kg) satellite to a Sun-synchronous orbit, one which offered customers a price of $22,675/lb ($50,000/kg) of payload. That was almost half the Stargazer/Pegasus price. Sun-synchronous orbits are inclined close to 90° to the planet's equator and pass over both poles so that they maintain the same relative position of the Sun at any given spot on Earth. These orbits are mostly for Earth observation and for the monitoring of environmental conditions from relatively low altitude and always below 620 miles (1,000km).

The performance requirement for the LauncherOne rocket to be carried by White Knight Two was determined by the market survey for potential customers. With technology advances offering increasing miniaturisation, small satellites would suffice for such specialised roles and in the niche markets they would fill there was very little commercial competition, other than the air-launched Pegasus

STARGAZER

N140SC

BELOW • Servicing the small satellite market, a Pegasus rocket acquires its payload in preparation for an air-launched flight into space. (NASA)

RIGHT • The nanosatellite market has grown exponentially within the past 15 years raising expectations of an expanding commercial marketplace. (Erikkulu/ wikicommons)

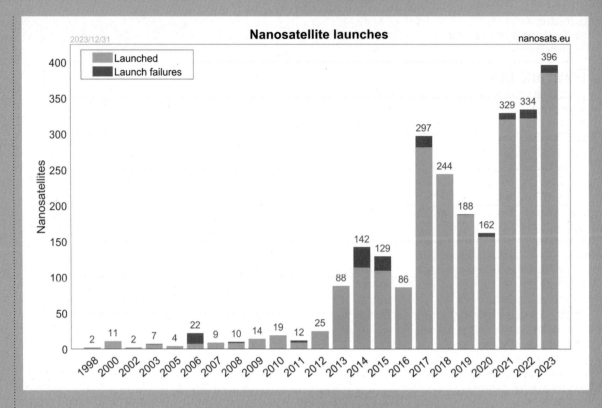

Nanosatellite launches

2023/12/31   nanosats.eu

Legend: Launched / Launch failures

rocket which, being older technology, could lose business to the more refined LauncherOne rocket. Key to that was the cost and performance of the two-stage design and initially a thrust of 47,000lb (209kN) was selected for the first stage, named NewtonTwo, with a smaller-scale version delivering a thrust of 3,600lb (16kN) named NewtonOne for the second stage.

The prototype test engines were ready by 2014 and NewtonOne was operated successfully for a full-duration run of five minutes while NewtonTwo completed some short-duration runs. But they would not be the motors finally employed for LauncherOne. Virgin Galactic acquired a facility at Long Beach, California, for the manufacture of rockets as development of the NewtonThree and NewtonFour motors got under way. The company reassured potential customers that it could begin flying satellites in 2016. But in the need to match weight-lifting requirements, the size and mass of the LauncherOne rocket grew beyond the carrying capacity of White Knight Two. A new carrier-plane would be required.

The business case for the project was consolidated on June 25, 2015, when the prospective satellite broadband provider, OneWeb, contracted with Virgin Galactic to launch 39 satellites with options on a further 100. By this date LauncherOne had grown in size and with longer fuel tanks came an extended burn duration. The liquid propellant first stage would now be powered by NewtonThree, delivering a maximum thrust of 73,500lb (326.9kN) for 180 seconds, the second stage powered by a liquid propellant NewtonFour motor providing a thrust of 5,000lb (22.24kN) for a total of 360 seconds. A second firing would circularise the satellite's orbit.

The search for a replacement aircraft began at the end of 2015. For Virgin Galactic the choice was obvious, a retired Boeing 747-400 from Virgin Atlantic, bought outright from Boeing at the expiration of the lease. By now, with modifications required to the 747-400 to equip it with a shoe pylon on the underside of the port wing inboard of the engines, OneWeb was told that operations had drifted into 2017.

With a total take-off weight of 550,000lb (249,480kg) including the 57,000lb (25,855kg) rocket, it would be inevitable that the converted jumbo jet would be named *Cosmic Girl*. Initially, flights would take place from Mojave Air and Space Port, with *Cosmic Girl* flying out over the Pacific Ocean to release LauncherOne for flight. There were other locations around the world from where flights could begin.

With a length of 70ft (21.3m), the first stage had a diameter of 6ft (1.82m) and would push the configuration to a speed of 8,000mph (12,872kph) while the second stage, with a diameter of 4.91ft (1.49m) would accelerate itself and the payload to an orbital speed of 17,500mph (28,258kph). In the revised configuration, LauncherOne could place a 1,100lb (499kg) satellite in a low Earth orbit or up to 661lb (300kg) to a Sun-synchronous path.

On March 2, 2017, Virgin Galactic moved its LauncherOne programme to Virgin Orbit specifically to handle operations, marketing, and flight support, and for defence and national security customers it set up Vox Space. Delays inevitably set in and not before August 2018 did test and evaluation flights commence with *Cosmic Girl*, a year in which OneWeb fell into financial problems, cancelling all but four flights. It would file for bankruptcy in 2020 before receiving financial

BELOW • Stratolaunch based its own market potential on the giant Model 351 from Scaled Composites in a configuration similar to that of the much smaller White Knight Two. (Scaled Composites)

guarantees and a modest bailout and re-establishing itself with new investors. None of its satellites ever flew on LauncherOne.

Down and Out

Virgin Orbit's competitor Stratolaunch rolled out its giant six-engine aircraft (N351SL), informally named *Roc*, in May 2017 and it took to the skies for the first time on April 13, 2019. A month later the company halted operations and put its aircraft and all its assets up for sale. With a maximum take-off weight of 1,300,000lb (589,680kg) carrying a 550,000lb (249,480kg) payload, it was similar in appearance to Scaled Composite's White Knight Two. With great potential for carrying oversized loads suspended beneath the inter-wing structure between the two fuselage sections, it was taken over by Cerberus Capital management and by early 2024 it had conducted only 13 flights.

The first active flight of LauncherOne took place on May 25, 2020, but a technical problem in the first stage shut it down and the rocket fell into the Pacific Ocean. A second flight on January 17, 2021 put 10 nanosatellites with a total weight of 253lb (114.7kg), in orbit and well below its capability. These were CubeSats for a NASA educational programme, and a range of US universities, secured in orbit with a successful second firing of the second stage. A second successful flight was logged on June 30, 2021, with third and fourth launches from *Cosmic Girl* in January and July, 2022. Most of the payloads for these five flights were for US government programmes, including NASA and the Department of Defense. But there was little interest from the commercial sector.

Development of the LauncherOne capability cost more than $1.1bn and the market to recoup that investment was fragile at best. To attract international customers, Virgin Orbit intended to launch satellites from several airports around the world, including an interest in servicing flights from Newquay in Cornwall, England. Seeking to place the UK at the forefront of the new age of space commercialisation, on September 30, 2022, the government invested in Spaceport Cornwall from which to operate *Cosmic Girl* and LauncherOne flights.

To host those activities, Virgin Orbit UK Ltd was formed out of the parent company and the British government contributed £7.5m for upgrades to the airport, re-badged as Spaceport Cornwall for that side of its business, opening on September 30, 2022, and gaining an operating licence from the Civil Aviation Authority on November 16.

The first of a series of planned flights took place on January 9, 2023, when several thousand people converged on Newquay to watch the late evening take-off. Carrying nine satellites, the rocket was released from *Cosmic Girl* off the south coast of Ireland and began its flight to

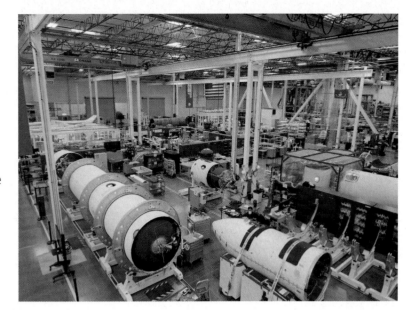

ABOVE • *LauncherOne rockets on the assembly line. Note the composites preparation bays at extreme rear in this image. (Virgin Orbit)*

space, the first stage performing as planned before the second stage failed to complete its burn when a clogged fuel filter starved the motor of propellant, the inert stage and its payload falling back into the sea. Elated initially, despondency descended over the night-time launch site for what was intended to be an inaugural event supported by the UK Space Agency and local government.

Virgin Orbit had many plans, including a three-stage variant that was intended to support very small spacecraft sent to Venus, while the Polish Space Agency had booked a launch for a national satellite it planned to send into orbit. But it was insufficient business to sustain a viable programme, which had already cost many times the original estimate. Virgin Orbit halted operations in March 2023 and filed for bankruptcy on April 4. Its assets were sold off to other aerospace companies. Spaceport Newquay leaves open opportunities for a return to space-related business in the future but with a different operator.

By early 2024 the commercial endeavours of Richard Branson had seen a planned halt to fee-paying space trips to the edge of space while a new generation of rocket-ships was developed and his ambition to challenge the orbital launch market addressed by Orbital Sciences and Pegasus had collapsed. But the challenge of space commercialisation brought high risk and small returns, until a much more ambitious approach ushered in a transformation that would shake the foundations of the space-faring world. But not before others had tried to turn rocket science into commercial success.

BELOW LEFT • *Cosmic Girl (N744VO) drops a LauncherOne two-stage rocket, sending it on its way into orbit. (Virgin Orbit)*

BELOW RIGHT • *Modest in its frontage, Newquay Airport nursed ambitious plans to be the UK's space port for sending small satellites into space. (Christian Lee)*

FROM OLD SPACE TO NEW SPACE

ABOVE • In the mid-1980s, the Reagan administration launched the Strategic Defense Initiative, or 'Star Wars' ballistic missile defence programme, which envisaged orbiting battle stations for which McDonnell Douglas proposed a commercial launch system. (SDIO)

RIGHT • A legacy of Old Space and now retired, one of the most powerful expendable rockets developed by the United States, a Delta IV with strap-on boosters thunders away from the launch pad. (USAF)

NASA experienced setbacks to its human spaceflight programme in the catastrophic loss of Space Shuttle *Columbia* when it broke up on re-entry on February 1, 2003. Not before July 26, 2005 had the Shuttle returned to flight with three orbiters instead of four.

Several years of deliberation over future plans pitched the government agency toward a future based on deep-space exploration of the Moon and Mars, leaving support for the International Space Station in the hands of an emerging commercial market for launch vehicles and spacecraft capable of operations into low Earth orbit. After the last Shuttle flight in 2011, the market began to open at a surprisingly buoyant pace.

New entrants were emerging - Elon Musk with SpaceX and Jeff Bezos with Blue Origin - each competing for different market sectors. Blue Origin was fast becoming a credible competitor to Virgin Galactic. Soon to be offering ballistic flights to the edge of space and back, Bezos chose development of a more conventional capsule launched on a reusable rocket. By 2015 the frontier was wide open and with increasing call on investors and a new cadre of entrepreneurs. But just how did that come about?

In the post-Shuttle era, where the big space station in the sky required regular supplies of food, water, and science

After the Shuttle – what next?

All these were on the point of being cancelled in favour of moving satellites and spacecraft across to the Shuttle, but when *Challenger* was destroyed in January 1986, production of these expendable rockets resumed. Only the US Air Force had insisted on the Titan being preserved throughout the shift to the Shuttle, as a back-up system which was sorely needed in the following years. Atlas and Titan served to launch government payloads for the US Air Force, NASA, and other organisations, but the workhorse for launching commercial satellites was the Delta vehicle which attracted launch contracts on a reimbursable basis from many operators around the world.

With a redesigned and stretched main core stage and more powerful upper stage and solid propellant boosters, the final generation of Delta II could lift 10,960lb (4,971kg) into orbit, the last of those flying as recently as September 15, 2018, logging 155 launches with only two failures. The first Delta II had flown in 1989 but since 1960 preceding Delta launch vehicles had been used to send into space a wide range of satellites for commercial customers and for governments around the world. It was the prolific array of payloads booked aboard these launch vehicles that stimulated interest in commercialising the industry.

There is a postscript to the Delta story. Completely redesigned but bearing the same name, the Delta IV series offered customers the option of several different

BELOW • Last in a long line of Atlas launch vehicles, this derivative variant launched a satellite for the National Reconnaissance Office, a frequent-flyer on Atlas and a customer seeking cheaper tickets from commercial competitors. (NRO)

payloads, the Orbital Sciences Cygnus logistics supply spacecraft launched on an Antares rocket was already backing up SpaceX and its Dragon cargo capsule. And there were more on the way from countries as far afield as Canada, New Zealand, and the United Kingdom.

The fuel for this surge in entrepreneurial activity was a buoyant satellite market and real progress in lowering the cost of access to space. But that had been a long time coming and hard to achieve. And it had fostered a revolution across several different competing platforms, all aiming to reduce the cost of launching people and cargo into space, considered the key to making orbital flight accessible to a wider range of people, companies and organisations. But it would come from the private sector, encouraged by government money supporting agencies desperate to offload expensive projects away from the public purse.

The cost of space transportation has been covered in previous chapters, where the initial array of available launch vehicles came out of military missiles developed by the army and the air force. The US Army produced the Redstone, from which came the Juno launcher, and Jupiter missiles which were adapted for a brief period sending some early satellites into space. The US Air Force had produced the Thor, which became the Delta launch vehicle, and Atlas and Titan ICBMs which saw considerable adaptation into space launch vehicles.

the development cost. In the rockets which were being used, all that had been paid for when they had been developed for military purposes with government funds. But a commercial rocket would have to pay back development in the price charged and a really efficient way of getting into space was necessary to do that and support a viable business model.

Theoretically, a SSTO does away with upper rocket motors to reach space and can get into orbit on the thrust of a single stage. This is extraordinary difficult to achieve because the amount of energy liberated by the most efficient known propellants is only marginally sufficient to deliver the power to overcome gravity and lift the weight of the rocket built of known materials. The efficiency of a rocket motor is defined by specific impulse (Isp), a calculation which displays the energy released by two chemicals – a fuel and an oxidiser for combustion. Liquid propellants offer a higher Isp, therefore greater efficiency, than solid propellants.

The most efficient commonly used liquid propellant is liquid hydrogen mixed with liquid oxygen, while other combinations such as kerosene (paraffin) with liquid oxygen are less efficient, but still release more energy than solid propellants, which are a binding compound of both fuel and oxidiser. The use of hydrogen and oxygen was proposed early in the 20th century by the Russian writer and mathematics teacher, Konstantin Tsiolkovsky. But liquid hydrogen is difficult to handle and requires insulated tanks to maintain it close to its boiling point where it converts from a liquid to a gas, around -435°F (-259°C) compared with that for liquid oxygen, at -297°F (-183°C). But the performance advantages are significant.

For a SSTO concept to work, it had to use the liquid hydrogen/liquid oxygen combination to achieve maximum efficiency from the propulsion system and an ultra-lightweight structure to support the necessary tanks, engines, and payload. And it had to be contained in liquid form because that is 800 times denser than gaseous oxygen and 850 times for hydrogen, which means it requires that much less volume to contain it. Such propellants carried at super-cold temperatures are known as cryogenics.

Although the Russian theoretician Tsiolkovsky worked out the equation for these propellants and proved they were the most efficient, it was the Americans who made the first great strides in producing liquid oxygen/liquid hydrogen rockets which made each unit of thrust do more work and lift heavier payloads. Which is why the Saturn V, with cryogenic

ABOVE • Legacy heavy-lifters will be around for a few more years, this Atlas photographed before sending NASA's Curiosity rover to Mars in 2011. (NASA)

configurations with strap-on boosters and upper stages, the Delta IV Heavy providing a payload capacity of up to 63,470lb (25,290kg), although this option was for a very specific type of low Earth-orbit payload. First launched on November 20, 2002, the Delta IV was retired after its final flight in April 2024 by when it had completed 45 launches, all but one a total success. In all, there had been 389 flights bearing the name Delta across 64 years of service.

As of 2024, all but the Old Space Atlas derivative launch vehicles have now been retired, their heavy costs forcing high prices which have brought about the emergence of New Space launch vehicles. These have emerged from commercial companies outside government ownership and subject only to the regulatory framework set down by the Federal Aviation Administration. It had been a long and tortuous path for the private sector to produce the organisation, finance, innovative designs, and entrepreneurship to provide what is now a New Space age of low-cost space transportation with partially reusable launch vehicles. How did that happen?

Rocket Science

By the late 1980s, many rocket scientists and space engineers in the United States were looking to develop a commercially viable launch vehicle to replace the existing inventory of expendables. It naturally fell to those companies already providing hardware for the US space programme to turn their talents to such an enterprise. The key to cheaper space transportation was a single-stage-to-orbit (SSTO) vehicle which could be returned for reuse and thereby offer lower operating costs. Most of the price charged for launch services did not incorporate a portion of

RIGHT • An early proponent of cryogenic propulsion and all things related to space and human occupation of orbital stations, the Russian Konstantin Tsiolkovsky inspired first-generation rocketeers. (Author's archive)

propellants for its second and third stages, developed to send Apollo spacecraft to the Moon, was more efficient as a launch vehicle than more recent rockets such as Starship from SpaceX which have greater thrust but less payload capability.

The idea of using cryogenic propellants and a lightweight structure for a SSTO capability had early advocates, nurturing in many a gnawing desire to realise the dream of launching into space by a single vehicle. It had been the stuff of science fiction and was frequently coupled with a vertical-take-off-and-landing (VTOL) capability, one in which the SSTO not only makes it into orbit but comes back and lands vertically on legs. From the 1960s there were many proposals for such a VTOL/SSTO spaceship capable of routine access to space. And even if the unit costs with such a system were high, repeated use of the same vehicle could significantly lower the cost of each flight.

Philip Bono, an aerospace engineer at Douglas Aircraft during the 1960s was an early proponent of these concepts and presented several proposals which were influential at the time. His work was publicised by the British engineer and writer Kenneth Gatland. When the company merged with McDonnell, Bono drafted variants of the S-IVB third stage of Saturn V for such an application and these designs encouraged others to propose similar concepts. On through the 1970s, several SSTO concepts were proposed as alternatives to the Space Shuttle and in the late 1980s the famed aerospace engineer Max Hunter presented a SSTO transportation concept to his employer, Lockheed Martin, which the company rejected as too expensive and too speculative.

In the aftermath of the loss of Space Shuttle *Challenger* on January 28, 1986, the US military became concerned that a rapid and frequent launch service would not be available to support its Strategic Defense Initiative (SDI), a plan to deploy laser and directed-energy weapons in space to shoot down any incoming missiles and warheads threatening the United States. Dubbed 'Star Wars' by the press after the film of that name, SDI envisaged multiple launches at routine schedules to establish and support constellations of low-cost satellites and battle stations to set up and maintain what the Reagan administration called the Peace Shield.

After *Challenger* temporarily grounded the Shuttle, the Department of Defense asked the Aerospace Corporation, a civilian arm of the US Air Force, and the High Frontier lobby organisation for SDI, to analyse these concepts. There was concern about the technical viability of the SSTO concept and the marginal plausibility of the technology. But Hunter had a novel idea to get funding from the government and on February 15, 1989, he met with Vice President Dan Quayle to propose a competitive programme to provide a deep-dive into the feasibility of such a concept. A request for study proposals was issued in August 1990 and four companies responded with McDonnell Douglas, Rockwell, General Dynamics, and Boeing receiving contracts for that work.

Front-runner McDonnell Douglas worked to a fast-turnaround/fully-reusable objective and recognising the highly experimental nature of the concept, Hunter designed a simple, small-scale prototype as a concept-demonstrator called DC-X with the full-scale production version known as

ABOVE • *The high launch model projected as necessary to support the anti-missile screen involved ground-based and orbital elements, a demand revitalising the search for reusable launchers, preferably operated by private companies. (SDIO)*

ABOVE • President Reagan and Soviet Premier Mikhail Gorbachev meet in Geneva as the Cold War comes to an end, redirecting national priorities and challenging NASA to adjust to lower budgets. (The White House)

the Delta Clipper. The SDI programme wanted a suborbital test vehicle capable of reaching a height of 284mls (457km) carrying a 3,000lb (1,361kg) load demonstrating precise vertical landing back at the launch site and a return to flight less than week later.

Clippers for Space?

DC-X would have a height of 39ft (11.88m), a base diameter of 13ft (3.9m), an empty (dry) mass of 20,100lb (9,117kg) and a loaded weight of 41,700lb (18,915kg). Propulsion was provided by four Aerojet RL-10A-5 cryogenic motors delivering a thrust of 14,500lb (64.5kN), a legacy from the Centaur, variants of which had been carried as the upper stage for Atlas and Titan launch vehicles. They could be throttled between 30% and full thrust, essential for controlling the final stages of landing and touchdown at slow speed. With four gaseous-hydrogen attitude control thrusters, the primary structure was formed from a graphite epoxy composite with a silicon thermal protection.

A significant achievement in the design was the fully automated launch and flight control system, pre-programmed for a set profile calling for only three personnel in the control bunker to take it through a full flight, one of those to support ground activity. This presented a completely different approach and is one of the legacy benefits from the programme, instilling a new approach to launch support and control, a far cry from the 450 people required to get an Apollo Saturn V off the pad from the Kennedy Space Center. The price and availability of commercial-off-the-shelf (COTS) components also contributed to significant cost savings as computer chips are cheaper to employ than people.

The first flight of DC-X, which would be named Clipper, took place on August 8, 1993, and was a successful demonstration of technique when it reached an altitude of 151ft (46m) and returned to the ground 59 seconds after lift-off. However, after demonstrating that it could move laterally and return, a small fire broke out on landing due to heat building up on the aft plate section. All DC-X flights took place from White Sands Space Harbor, New Mexico, the place where much early rocket development and testing in the 1950s began.

RIGHT • Delta Clipper pioneered vertical launch and landing for a rocket stage which McDonnell Douglas hoped would significantly lower flight costs through a commercial enterprise. (NASA)

That first flight was very nearly postponed when a seriously heavy storm all but waterlogged the entire area, causing personnel to rush around bailing out water-filled trenches. But as proof that it could be achieved, a rocket had been made to land vertically back on Earth for the first time. In that regard it was a resounding success. The next flight occurred on September 11, a 66 second flight in which DC-X reached 302ft (92m), followed on September 30 by the third lift-off to a height

of 1,210ft (370m) during which it completed a 180° roll before landing 73 seconds later back at the launch pad.

But the bigger picture of which DC-X was just a minor player changed significantly when the Department of Defense pulled the plug on Star Wars, eliminating the requirement which could have had significant applications right across the space sector. Under new funding from NASA and the Advanced Research Projects Agency (ARPA), flights continued as a research programme. The next flight under that rebadged effort took place on June 20, 1994 when DC-X reached a height of 2,850ft (870m) in a flight lasting two minutes 16sec. The next launch seven days later was a partial failure due to a small explosion during launch, but it reached 2,590ft (790m) and lasted 78 seconds. Three more flights took place between May 16 and July 7, one of which logged a record height of 6,600ft (2,911m) and a duration of two minutes 12sec before the basic DC-X programme again changed ownership and direction.

From late 1995, NASA took over the Delta-X Clipper and made a number of hardware modifications and improvements, changing some materials in the fabrication of tanks and the structure and adding upgrades to some existing systems. The refurbished vehicle was badged DC-XA and made its first flight as such on May 18, 1996, in which it reached 801ft (244m) before landing 62 seconds later but suffering a fire in the aeroshell due to the slow descent. Three more flights occurred between June 7 and July 31, 1996, during which it achieved a maximum height of 10,300ft (3,139m) and a duration of two minutes 22sec. Unfortunately, it was destroyed on its final landing when one of the legs failed to deploy and it toppled over.

The cause of that failure was attributed to human error during pre-flight preparations, due, it was officially recognised, to an exceptionally fast-paced and highly labour-intensive effort where the human workload was raised to the point where errors were beginning to set in. Engineers working the programme raised deep concerns over the continuous changes to flight profiles, test schedules and the abundance of paperwork involved at NASA in what was evolving into a top-heavy bureaucracy. Where once it had been a breeding ground for ingenious ideas leading to spontaneous and highly creative research and development, NASA had become a burdened agency and a victim of post-Cold War uncertainties.

The dual effect of the January 1986 *Challenger* disaster and the collapse of the Soviet Union less than six years

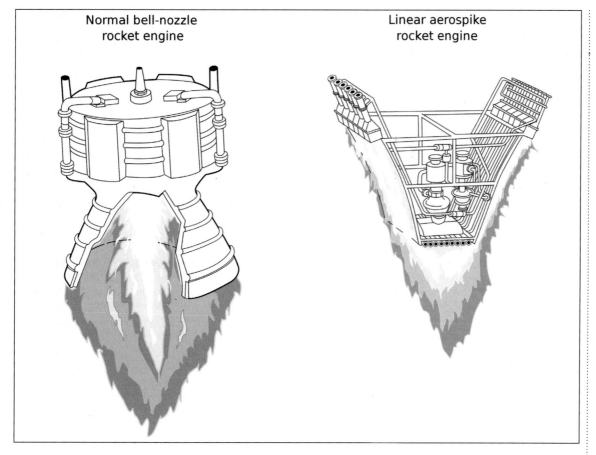

Normal bell-nozzle rocket engine

Linear aerospike rocket engine

later left NASA floundering for solutions: to a far better and more efficient way to gain access to space, and to its real purpose now that it had achieved measurable gains over its former protagonist. That competition, the very reason why NASA had been formed, had now evaporated into the mists of uncertain foreign policy strategies. NASA's new role in a very different world was cemented as playing lead-agency for a wide range of international programmes, epitomised by the introduction of the new Russia onto the International Space Station as a contributing partner.

Everyone at NASA knew that support for the station required a better and more effective crew supply and cargo delivery system than the Shuttle could provide. The Shuttle was expensive, lifting cargo at a cost of $24,500/lb ($54,000/kg) and the government wanted to see that reduced by a more efficient and reusable system to less than the cost of a Titan IV launch, at $11,200/lb ($24,700/kg). That could only be achieved by a radical change in the way operations were conducted, with return and reusability a key to lower costs, far lower than the Shuttle could deliver.

The search for a solution began as early as December 1993, when NASA launched its Access to Space study to find a way of reducing transport costs by 50%, increasing safety levels and enhancing operational sustainability. Three options were considered, including upgrades to the Shuttle to keep if flying through 2030, developing a new expendable launcher and starting a transition away from the Shuttle in 2005, and development of a new and reusable system to begin the transition in 2008. A common mission model was applied to consideration of all three, much as had been done in 1970-1972 to determine the kind of Shuttle to build. That had been flawed by an over-ambitious estimate of the number of annual flights which could be achieved and the demand for Shuttle custom, which had driven it to be cited as the replacement for all expendable launchers.

The study recognised that there was a culture-change in launch vehicle development in which technology could be applied to robust operational margins rather than sheer performance. The mission model showed that 90% of all space cargo for low orbit would weigh less than 20,000lb (9,092kg) and be less than 20ft (6.1m) in length, with an estimated 39 flights each year. That figure was considerably less than the 60 flights per year estimated for the Shuttle when Mathematica had conducted their Shuttle study almost 25 years previously. Moreover, it recommended development of a fully reusable SSTO vehicle as top priority.

At this time, the DC-X Clipper was undergoing flight tests, and it became an integral part of the development programme for a reusable SSTO. As a direct result of the

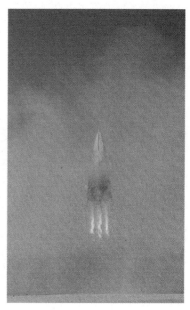

study and as a determination at the highest level up to the White House, a national space transportation capability was inaugurated and two other technology demonstration projects were started in 1994: the X-33 and the X-34, each with a very different design concept.

A New Venture

The X-33 is of special interest because it emerged from the winning bidder Lockheed Martin as a demonstrator which the company wanted to develop into a fully commercial launch system owned and operated by them and leased to NASA for services. McDonnell Douglas and Rockwell International had also bid for the work under a request for proposals issued in 1994, but they had little interest in owning the result. It did, however, come at the same time that Orbital Sciences got a contract to develop the X-34, which was an air-launched, hypersonic research project and quite different to the X-33.

By this time, NASA itself questioned the role of trucking operator which it had slipped into with the Shuttle, which had been conceived back in the late 1960s as a supply vehicle for a wide range of other activities. These included operating a large space station, flying back and forth to lunar bases, operating a space tug for moving satellites around for repair and refurbishment in space and using a nuclear upper stage for manned flights to Mars. All elements other than the Shuttle had not been possible due to the collapse of NASA's budget, leaving the agency with only unmanned spacecraft to explore the solar system and the ISS to run.

Now, the idea of offloading station supply and cargo delivery flights to a private owner/operator was highly

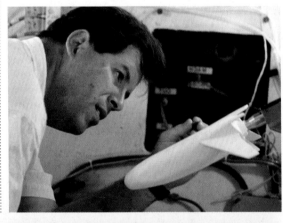

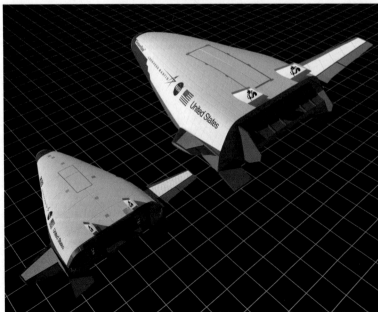

attractive, and it was the start of a lengthy search for such a company, encouraged by government development money to start the process. It was in that vein that the X-33 was funded by both NASA and Lockheed Martin in the hope that the national space agency could be free to find money thus saved for its core goal of exploration and for extending human space flight back to the Moon and on to Mars. It was awareness of that dream that stimulated other entrepreneurs, wealthy billionaires who would start afresh and build new companies and very different ways of operating which would, in turn, liberate NASA for its core purpose.

From the outset, the X-33 was a suborbital demonstration project with a requirement to reach an altitude of 49 miles (78.5km) and around 9,600mph (15,446kph) in a series of 15 flights. It was to be powered by a unique form of rocket motor called an aerospike engine. Instead of having a cone-shaped exhaust nozzle it would distribute the exhaust by firing along one side of a wedge-shaped structure, constrained on its 'outer' surface by the pressure of the atmosphere. As the vehicle ascended, the pressure on the 'virtual' side of the motor would automatically shape the exhaust as the pressure declined with increasing altitude.

The expansion effect that occurs with increasing altitude can be seen when watching a rocket ascending from the pad, its exhaust plume gradually changing from a downward direction to an expanding, bell-shaped plume as it climbs into the rarefied atmosphere. As noted earlier, a fixed, bell-shaped exhaust nozzle can only ever be optimised for one part of the flight path, but an aerospike rocket motor allows that adjustment to take place naturally. It can come in two forms, either with a plug-shaped nozzle in which the exhaust flows around and down the exterior surface of an inverted cone, or as a linear motor as selected for the X-33.

The two XRS-2000 linear aerospike engines for the X-33 had a total thrust of 410,000lb (1,823kN). With a length of 69ft (22m) and a width of 77ft (23m), the X-33 had a projected weight of 285,000lb (129,276kg). Lockheed Martin calculated that with ultra-lightweight tanks, the propellant mass fraction (or the quantity consumed in reaching orbit) of an orbital successor to X-33 could achieve 0.9. The higher the mass fraction the better. For the Shuttle it was only 0.79. That successor was marketed by Lockheed Martin as the VentureStar.

With a height of 127ft (38.7m) and a span across the wings of 128ft (39m), VentureStar would have had a lift-off weight of 2,200,000lb (997,900kg) with propulsion provided by seven XRS-2000 linear aerospike engines producing a total thrust of 3,010,000lb (13,388kN) carrying a 44,000lb (19,958kg) payload into orbit. Like X-33, it would launch vertically and return from space to

land like an aircraft on a paved runway. The technology required to provide that capability was highly advanced and the materials and engineering were a challenge to the national resources of NASA and Lockheed Martin. So much so that the X-33 and VentureStar were cancelled in 2001.

Contracted to Orbital Sciences initially as a prototype SSTO technology development vehicle, the X-34 was repurposed in June 1996 into a platform for evaluating the necessary engineering steps essential to achieving that broader goal. With a length of 58.3ft (17.77m) and a wing span of 11.5ft (3.5m), it had a conventional configuration and was designed to be air-launched off an adapted L-1011 TriStar. With a mass of 18,000lb (8,200kg), it had a Fastrac engine developed at NASA's Marshall Space Flight Center producing a thrust of 60,000lb (266.8kN) and was capable of achieving Mach 8. As a pilotless research aircraft, it would have reached an altitude of 50 miles (80km) and

returned through the atmosphere for a fully automated landing on a conventional runway.

Two X-34s were built by Orbital Sciences, an entrepreneurial company founded by three graduates from Harvard Business School and well known for having developed and operated the Pegasus air-launched satellite launcher. Funded by NASA and a number of other government agencies, the X-34 fell victim to sustained year-on-year reductions to NASA's budget and the programme was cancelled in March 2001, along with the X-33 and VentureStar. Research on these separate programmes, including the earlier test flights with Delta Clipper, provided some valuable information, but none resulted in operational vehicles.

Throughout the 1990s, NASA had struggled to find a cheaper way to launch payloads into space and largely failed. By sticking to traditional ways of cutting costs, through reusability and greater efficiency, it had run out of accessible technology and the money to sustain efforts that challenged conventional solutions. What was required would turn the industry on its head and transform the marketplace, opening up space to a new generation of owners, users, and operators. And that would come not from government agencies and big aerospace corporations, but from men with very deep pockets creating a completely new industry. Out went evolution – a revolution was about to begin.

SUBSCRIBE TODAY!

Air International has established an unrivalled reputation for authoritative reporting across the full spectrum of aviation subjects.

ELON MUSK'S
FLYING
FALCONS

Enter the billionaires club

As noted in previous chapters, there have been many attempts to reduce the cost of launching satellites and spacecraft by means of a single stage capable of carrying payloads into orbit and returning to either a vertical or a horizontal landing. In all concepts proposed, the underlying goal was reusability, and the only way to do that was to recover the entire structure and launch it again with a fresh payload, slashing the price to customers. Using an aviation analogy, the single use was like building a Boeing 747, flying it from London to New York, and throwing it in the Hudson River after that one flight.

Had aircraft not been reusable, there would have been no aviation industry. Strapped for cash as its budget plummeted, NASA was unable to fund extravagant and revolutionary ideas to achieve that for space transportation. Yet, despite having launched unmanned robots to most planets in the solar system and put humans on the Moon, although reusability was the key to a space-based economy it was the most difficult to achieve. Not until it had been made possible could launch costs come down and flights to space become routine at knockdown prices.

Enter Elon Musk, a man in a hurry with a largesse of ideas to match the size of his bank balance and the sheer audacity to tear up the rulebook and offend traditionalists with his outrageous concepts. Musk wanted to give the world a reusable launch system by turning the general laws governing rocket science on their head and declaring an answer to find the question: that by returning rocket

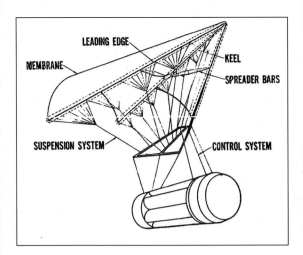

stages back down after they had finished their work, they could be used again and shatter the profits for big launch vehicle providers, upsetting the business model for major aerospace corporations and national space agencies alike. The question: how can you do that when nobody else had been able to return them to Earth?

From the very beginning of the space programme, people had tried to do just that. In the 1950s, the brain behind the Saturn rockets that would take men to the Moon, Wernher von Braun had wanted to fly more than a hundred Saturn rockets each year, build a space programme supporting several space stations, and provide a network of Moon stations and surface bases while launching expeditions to Mars. To do that he needed to reuse those first stages, accepting that the upper stages, smaller and cheaper, could be expendable. The big money was spent on the giant first stages, boosters that if recovered could significantly cut the cost of an overall programme.

The simplest way was to recover those stages from the sea into which they would fall after separating from the upper elements of the launch vehicle. Von Braun's team studied the use of 105ft (32.5m) diameter parachutes deployed in the denser regions of the atmosphere to slow them down, and solid-propellant retro-rockets fired just above the surface to bring them to a gentle splashdown. Afloat due to the buoyancy from empty tanks, they would be recovered and towed 200 miles (322km) back to Port Canaveral where they would be cleaned and used again. So committed was the von Braun team that Saturn I was manufactured using a 5456 aluminium alloy which welds and repairs easily but at the expense of lower strength-to-weight ratios than other alloys which corrode more on contact with water.

Other ways of returning rockets intact and for reuse involved parawings, inflatable lifting surfaces which, when deployed, could bring spent stages gently back to solid ground instead of having to suffer the effects of seawater. Von Braun was able to apply study results from wartime research on recovering V2 rocket stages this way when used for tests which did not carry warheads for military use. Because more than 700 V2 rockets were launched in the course of the missile's development, this approach made good sense. Von Braun wanted to apply the same logic to his Saturn series of launch vehicles which, when the studies into stage recovery really took hold, were anticipated to require large numbers of test flights.

It turned out that the average frequency of Saturn flights between October 1961 and July 1975 was 2.28 per annum for the 32 launches in those 14 years, compared with more than 250 V2 missiles per annum test-launched for the three years between the first in March 1942 and the last in January 1945. The expectations of the von Braun team were never

met, but for a time in 1959 two members of the rocket team considered privatising the Saturn rocket and redesigned it into a configuration which they believed to be marketable, but the slow pace of the early space age made such plans impracticable.

Musk had a variation on this theme which went against the trend. In the opening year of the new century, he turned to space exploration and the phenomenal amounts of money it cost to launch payloads into orbit. Frustrated by the lack of progress in lowering those costs, and critical of the stranglehold on free-wheeling enterprise and the cult of the entrepreneur held by the major rocket companies, he set about upending the conventional approach and building a reusable rocket that could be returned to Earth and flown again.

But rather than recovering spent stages by parachutes and retro-rockets, he conceived that those stages could be steered back to the launch site by preserving a proportion of the propellants so that the same rocket motors that had got it up there could be used to slow it down. And, by using updated technology and electronic control systems, steer the descending stage to a precise touchdown point. In this way, those stages could be re-flown, many times over.

There was a penalty of course, in that the outright performance of the initial stage would be slightly compromised by shutting down the rocket motors prior to depletion so that there would be sufficient propellant remaining to push it back around toward the launch site and then function as retro-rockets to lower it to a gentle touchdown. But that was OK. The significant reduction in the price to customers could attract frequent flyers

RIGHT • After the tragic loss of Columbia in 2003, NASA examined a wide range of modifications to the Shuttle but opted to retire it and leave resupply of the International Space Station to commercial operators. (NASA)

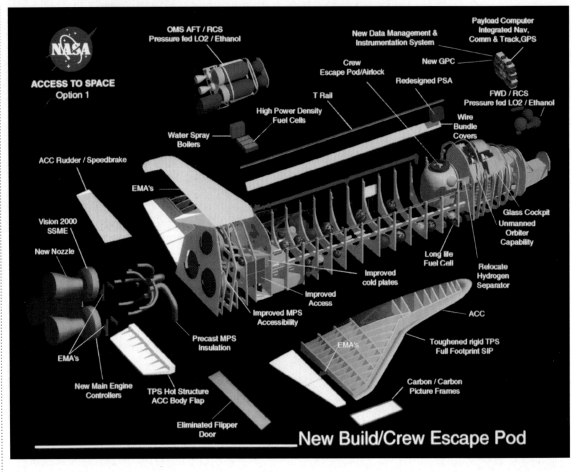

ACCESS TO SPACE
Option 1

OMS AFT / RCS
Pressure fed LO2 / Ethanol

New Data Management &
Instrumentation System

Payload Computer
Integrated Nav,
Comm & Track,GPS

Crew
Escape Pod/Airlock

New GPC

Redesigned PSA

T Rail

FWD / RCS
Pressure fed LO2 / Ethanol

High Power Density
Fuel Cells

Wire
Bundle
Covers

Water Spray
Boilers

ACC Rudder / Speedbrake

EMA's

Glass Cockpit
Unmanned
Orbiter
Capability

Vision 2000
SSME

Long life
Fuel Cell

Relocate
Hydrogen
Separator

New Nozzle

Improved
cold plates

Improved
Access

ACC

Improved MPS
Accessibility

EMA's

Toughened rigid TPS
Full Footprint SIP

Precast MPS
Insulation

EMA's

Carbon / Carbon
Picture Frames

New Main Engine
Controllers

TPS Hot Structure
ACC Body Flap

Eliminated Flipper
Door

New Build/Crew Escape Pod

BELOW LEFT • A disruptive game-changer who refused to accept failure and pushed hard to lower the cost of space transportation and redefine the way humans explore the Solar System. Elon Musk has been in the vanguard of New Space for more than 20 years. (USAF)

BELOW RIGHT • Gwynne Shotwell has managed SpaceX since 2002, developing it into a successful business through partially reusable rockets, cargo and crew delivery vehicles. (SpaceX)

so that the overall revenue from a fleet of such launchers would be greater than that obtained from a single-use system.

This was great in theory if it could be made to work. Nobody had tried this before and there were sceptics as well as detractors and outright critics claiming that Musk was talking rubbish, and that blue-sky thinking would doom any such attempt to do that. There was big corporate risk at stake – to the companies and organisations holding the key to the existing rockets and space launch systems who saw a potential threat to their monopoly, built on high prices and high profit margins.

As Musk got to work, the primary operator in the United States was United Launch Alliance (ULA). Formed in May 2005 as a joint venture between Boeing and Lockheed Martin it sought to rationalise the Delta and Atlas launch vehicles of which each company had respective ownership. Following several years of rising prices charged for government payloads

launched on these rockets, wrangling over access to a reliable and cost-effective service threatened to breach a political determination to cap the price charged by contractors.

To prevent a government takeover of their respective products, Boeing and Lockheed Martin agreed to form ULA and fend off political intervention. It worked, and although prices remained high there was now visibility that launch services were based on a fair and agreed price without the prospect of a monopoly holding the taxpayer to ransom. Not an ideal situation because there was in fact a very real monopoly, in that there were no domestic competitors for defence and national security payloads and only Delta had any effective civilian market to serve.

Musk saw in this an in-built lockout for competitors because ULA held the big hand when it came to providing reliable launch vehicles with proven track records, factors vital in getting customers. Civilian operators and non-government organisations insured their satellites and brokers wanted proof that the satellites would make it into orbit and survive to start operating. Buying insurance was usually essential for getting the money to buy the satellites from a manufacturer and pay the launch provider (ULA) for sending it into space. And the higher the rocket's reliability, or actuarial spread as insurers call it, the lower the premium. The government self-insured for its military and non-military satellites and spacecraft but only selected launch vehicles with a lengthy log of successes.

Banking Options

Musk was the latest in a long line of disgruntled entrepreneurs and rocket scientists frustrated by the lockout they faced with this classic Catch-22 situation: to recruit orders from potential customers the operator had to show reliability over a credible series of successful flights; to get the flights to do that required money which would only be available from banks satisfied that the launcher was a success. To break the mould required a lot of money up front for high-risk ventures which might or might not prove successful.

But Musk faced more competition than that in the United States. Another player in the global market for commercial launch services was Arianespace, formed out of the collapsed deal for the Europa launch system when Britain withdrew its Blue Streak first-stage element on April 27, 1973. Five months later the French government led a consortium of continental European countries which would develop the Ariane launch vehicle, and, in its wake, the European Space Agency (ESA) was formed. Ariane was a bold decision, coming seven years after NASA had signed the reimbursable launch agreement with foreign countries which appeared to cancel any advantage in an autonomous provider outside the United States.

Mocked at the time by US operators who claimed the Europeans were reinventing the wheel with yet another expendable rocket when it was believed the Shuttle would replace all those throw-away launchers, the first Ariane 1 was launched on December 24, 1979, 16 months before the first Shuttle flight. Almost immediately thereafter, ESA set up Arianespace, owned by ESA and the French national space agency, the world's first launch services company to market Ariane.

Being government owned, Arianespace was not a commercial company and the first launch for a customer occurred on September 10, 1982. After a successful development programme through a series of increasingly more capable Ariane variants, by 2004 Arianespace was launching more than half the world's satellite traffic going into the lucrative geostationary orbit where telecommunications and TV satellites were operating and introducing the world to satellite TV.

In any assessment of the potential competition, Musk was up against the commercial heft of United Launch Alliance and Arianespace but there were other countries with the capability to launch satellites for foreign customers on rockets developed for national programmes. Soviet interest in making its wide range of launch vehicles available for commercial customers began through the Interkosmos programme involving Comecon countries of the Soviet bloc during the Cold War.

ABOVE • Successive evolutions produced the Ariane 5, seen here launching the Webb Space Telescope for NASA in December 2021. (Arianespace)

Established in 1967, through client states Interkosmos would compensate the Russian government for flying instruments, experiments, even astronauts from communist countries on Russian rockets for the national prestige it would bring. Eventually, astronauts from outside the pro-Soviet bloc and favoured communist countries would also fly on Russian rockets, seats paid for by Afghanistan, France, India, Japan, and Austria, although as an international exercise in goodwill the Russians found themselves paying for Helen Sharman's flight representing the UK when funds dried up.

To expand its commercial opportunities, the Soviet government formed Glavkosmos in 1985 to oversee all non-military space activity but which quickly became a marketing organ for selling Russian space capabilities to foreign countries. It still exists under the Russian national space operator Roskosmos and is credited with having launched more than 200 satellites and payloads.

During the late 1980s, Glavkosmos signed a deal with the US Space Commerce Corporation to support and sell launches on Russian rockets to clients in the West but, suspicious that the Russians could gain access to advanced technology, the US State Department closed the doors on that opportunity, and it got nowhere. Neither did a post-Soviet venture known as Excalibur Almaz, founded in 2005 and with its registered office in the Isle of Man off the coast of England, offering flights around the Moon on retired Russian military spacecraft. It was with these prolific opportunities in mind that Musk initially thought to buy Russian rockets to start a privately-run launcher programme before deciding that this was fraught with difficulties, both financial and political.

Japan had an active commitment to a national space programme and launched its own satellite on February 11, 1970,

LEFT • Europe's search for independent launch services resulted in the successful Ariane rocket, this being an Ariane 3 with solid propellant strap-on boosters. (Arianespace)

a new generation into technical subjects in schools and universities, India followed China in turning to satellites and rockets, at first buying payloads and launch services from the United States and then developing its own domestic industry.

Formed in August 1969, not least inspired by the Apollo Moon missions, the Indian Space Research Organisation (ISRO) would slowly develop a range of satellites and launch vehicles. By 1980 it had become the seventh country after Russia, the United States, France, Japan, China, and the UK to launch its own satellite. Soon, following now established practice, it offered its launch vehicles for customers around the world and orbited its first commercial payload in May 1999, to date sending more than 430 separate satellites into space for operators in 36 countries. State-owned, the service is not privately owned but, like NASA's reimbursable service of the previous century, it is a viable way of recouping taxpayer investment.

Thus, it was that government agencies in Europe, India, Japan, China, and Russia offered launch services using their own domestically produced rockets. Some operators and major organisations mixed ride options and chose different launchers, a good example being Inmarsat, the International Maritime Satellite Organisation for relaying voice communications ship-to-shore and between vessels at sea. The first three Inmarsat satellites in 1976 were launched on American Delta rockets, the next three between 1981 and 1984 went up on Europe's Ariane, followed variously by Atlas from the US and Proton and Zenit rockets from Russia, an H-II from Japan, and Falcon 9s from SpaceX.

There had been several failed attempts to penetrate this market with a privately-funded space launch system. Founded by David Hannah in 1980, Space Services Incorporated had such aspirations but no success until it hired former Mercury astronaut Deke Slayton and opted to purchase retired Minuteman ICBM rockets and cluster solid propellant motors from the second stage in a design known as Conestoga. Several iterations of the baseline design were proposed, and three flights took place in 1981, 1982 and 1995, two of which failed.

Flights of Fancy

But what got Musk into this business? One in which it is said that the only way to make millions of dollars out of

ABOVE • China has a rapidly expanding launch vehicle development programme, including the flight of crewed vehicles such as Shenzhou 13 shown here. (CNSA/ ForrestXYC)

but not until May 17, 2011, did it place its first commercial payload, a South Korean communications satellite, in orbit. Several other flights followed but it has never been a particularly successful endeavour. These were flown on launch vehicles developed by government money for its own space programme but Space One is a private company that developed a multi-stage rocket to place satellites in orbit. It failed in its first attempt on March 13, 2024.

China had been developing its own satellites and launch vehicles for domestic use as the country expanded into high-tech activities, placing its first satellite in orbit on April 24, 1970. It began selling its launch services on Long March rockets on April 7, 1990, when it launched AsiaSat 1 for a company in Hong Kong. Built by the US Hughes company for Western Union, it had been launched by Shuttle in February 1984 as Westar 6 but the Payload Assist Module that should have pushed it into its geosynchronous orbit failed and it was recovered by another Shuttle mission later that year. Returned to Earth it was in the possession of the insurance underwriters and sold to AsiaSat who re-launched it on a Long March rocket.

Elon Musk was particularly attracted to competing with China in the market for Long March rockets and that target became an early driver in his attempt to develop a truly independent launcher for commercial marketing. He believed that he could develop a rocket with the same capabilities as the Long March family and undercut the prices China was offering. But China was not the sole government-run competitor in Asia.

Seizing on space technology as a stimulant for national prosperity in the high-tech sector and using rockets to encourage

RIGHT • Early attempts in the US to develop a commercial launch vehicle included the unsuccessful Conestoga rocket supported by Mercury astronaut Deke Slayton. (Eric Grabow)

rocket flight is to start with several billions of dollars and work down. The reason is not as self-evident as many have asserted and that maxim has more truth in it than might be expected. For it is not solely for profit that Elon Musk began a journey more than 20 years ago which overturned convention and stunned a self-assured market convinced of its own persuasive arguments, locking out newcomers and dismissive of entrepreneurial ideas from newbies.

Born in Pretoria to a wealthy South African family on July 28, 1971, Elon Reeve Musk moved to Canada in 1989 where he gained citizenship through his Canadian-born mother, graduating from Queen's University, Kingston, Canada, and later from the University of Pennsylvania in the United States from where he moved to California in 1995. With his younger brother Kimbal he founded a city-guide software company, acquired by Compaq four years later for $307m, using some of that money to start an internet bank which evolved into PayPal through a merger with Confinity. In 2002, eBay bought PayPal for $1.5bn.

In March 2002 Musk founded SpaceX after meeting Robert Zubrin, a founder of the Mars Society advocating the colonisation of the Red Planet with a radical transformation in the way the space programme was funded, managed, and run. Zubrin had a large and committed following and Musk became a strong and vocal supporter, donating $100,000 to the organisation and supporting its plan to send a greenhouse to Mars where research could be conducted into growing food to sustain colonists.

Energised by the enthusiasm for interplanetary travel enshrined within the Mars Society, Musk became a staunch advocate of making humans a multi-planet species. His argument rested on the need to protect humanity from extinction by ensuring that there was another place in the Solar System where civilization could grow and prosper should a global pandemic, nuclear war, or a catastrophic asteroid impact threaten to wipe out humans on Earth. It was only through significantly cutting the cost of space travel that Musk believed that objective could be met and that he could make that happen by adopting radical concepts through revolutionary technological leaps.

Musk became obsessed with vertical integration, a corporate management and ownership concept in which all the essential peripheral elements are within the control of the central organisation, eliminating modular profits attached to each part of the supply chain, cutting costs, and reducing prices to expand demand through an economy of scale. These were ideas to which government bodies were opposed because organisations such as NASA saw themselves as procuring services from a base of manufacturing and production that they did not themselves possess or have any inclination to acquire.

NASA had the ideas and the plans and knew what they needed to get what they wanted but were unable to 'own' the concept which was instead enabled by specialist industries with track-records in success and reliability. Hence, all but a very few satellites and spacecraft were bought in from established manufacturers which were then able to apply the capabilities acquired through NASA contracts to expand their in-house

BELOW • With the most beautiful launch site in the world, Japan launches its space rockets from Tanegashima Island where periods of the year are declared no-fly months to allow fishermen undisturbed access to the sea. (NASA)

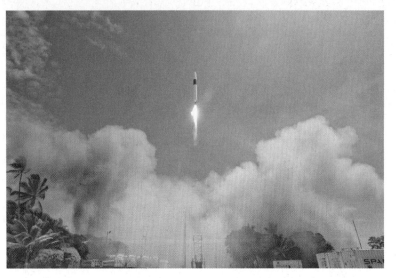

ABOVE • *On the final flight attempt of four, a Falcon 1 succeeds in reaching orbit on September 28, 2008. (SpaceX)*

BELOW • *Michael Griffin (left) meets Elon Musk (right) to discuss commercial potential in a New Space approach to fast-track development of rockets and spacecraft. (Author's archive)*

capabilities for producing commodities which they could sell to non-government customers. It was a form of capitalism that worked well for a sophisticated manufacturing industry because it raised the revenue for those specialist companies returning higher taxes on larger profits in a feedback loop. But it did nothing to overthrow the status quo and fuel ingenuity.

Musk believed he could use his wealth to slash launcher costs through a system of vertical integration and wholly-owned skill-sets which would all feed into the central purpose, increasing launch rates and revolutionising the space industry to support his dream of planetary emigration and settlement on other worlds. It would be a mistake to interpret Musk's intention as monetary gain and profit-making. For him, profit in this context fuels further advances toward a space-faring humanity which can only be funded through releasing revenue from investment and reinvesting money generated through a frequent-flyer rocket programme.

To achieve that, Musk tried unsuccessfully to buy retired Russian ballistic missiles to convert into satellite launchers but failed due to a lack of faith from the seller and their conclusion that, as he had no experience as a rocket scientist, he would fail and better to wait for a more qualified customer. Which never came. Nevertheless, Musk was recruiting support at home through the sheer audacity of his approach and the self-belief that attracted talented individuals from major aerospace contractors, many of whom were frustrated by corporate inertia.

Among these was the gifted and highly respected aerospace engineer Michael Griffin, who had accompanied

Musk to Moscow in the abortive attempt to buy rockets. Griffin would become NASA Administrator in 2005 and do much to prepare the agency for what was about to happen, arguably starting the process to save it from reneging on international commitments, explained in the next chapter. Musk's first employee was Thomas Mueller, another talented engineer who could bring radical concepts and make them work. A graduate in mechanical engineering and applied mathematics, Gwynne Shotwell joined SpaceX where she quickly rose to become president and chief operations officer.

As part of his vertical integration approach, Musk wanted to design and build his own rocket motors, something no major launch vehicle designer had done. While founding the Tesla car company he put his team to work on a motor he called Merlin to power the first stage of his Falcon 1 rocket, and another called Kestrel to power the second stage. With a thrust of 100,000lb (444.8kN), Merlin would have a burn duration of two minutes 49sec before the second stage would take over with a thrust of 7,000lb (31.13kN) and burn for six minutes 18sec. It would be capable of placing a 1,480lb (670kg) satellite in low orbit.

Musk's philosophy was to start small and build incrementally. His original plan to launch from a modified pad at Vandenberg Air Force Base, California, which was the West Coast launch complex for the US Air Force, became untenable when plans collided with existing flight schedules for other rockets from ULA. Instead, Falcon 1 would fly from Kwajalein Atoll and the first flight would carry a US Air Force test satellite for measuring space plasma. After numerous technical delays it was destroyed when the rocket exploded after launch on March 24, 2006. After technical delays, the second flight of a Falcon 1 on March 21, 2007, also failed, as did the third attempt on August 3, 2008.

Personal problems hit Musk in this period, with a divorce from his first wife, a failed round of financing for Tesla, and the last hardware procured for Falcon 1, dictating that if this one also blew up it would take his dream away with it. Musk asserts that this was the final throw of the dice and that if it rolled out of favour there would be no more money. Launch occurred on September 28, 2008, and it was a success, the first privately developed launch vehicle to reach orbit, albeit with a mass simulator named RatSat on this final trial flight.

Spacex Rising

Overall, the Falcon 1 programme cost Musk almost $100m and all of it out of his private bank account. But there was already a significantly greater prize ahead, one which could provide Musk with new opportunities so big that it could bankroll his dream of building giant super-rockets for taking humans to Mars, in very great numbers. Nevertheless, on the historic day that the world's first private rocket reached space Musk had only $30m remaining, which was insufficient for funding both SpaceX and Tesla for more than two months and there were no further contracts for his Falcon 1.

After a second successful launch on July 14, 2009, which placed a small Malaysian satellite in orbit, the Falcon 1 series was retired. "We couldn't get it to earn its place on the factory floor," said Gwynne Shotwell. SpaceX had wanted to develop a recoverable system for the first stage but that had not been achieved with this rocket and a successor was already well under way which offered greater promise, in payload capability and with the potential for recoverability. In Musk's mind, getting the expensive first stage elements back for subsequent reuse was a key factor in lowering transportation costs.

Any good product is defined by the market it addresses and so it was with SpaceX when it became apparent that income from the small satellite business addressed by Falcon 1 was just not viable. But a parallel development, Falcon 5 would incorporate five Merlin motors producing a first stage thrust of 357,440lb (1,590kN), about that of an early Atlas launcher and capable of lifting 9,300lb (4,200kg) to a low Earth orbit via a single Kestrel engine in the second stage. Merlin was seen as key to a progressively larger series of launchers, each building on the learning curve of predecessors.

Falcon 5 also had engine-out capability, retaining sufficient energy to reach orbit should a single motor fail after lift-off, the first since the Saturn series to have that. It was also the first SpaceX vehicle to have a planned stage recovery by parachute, much like the Shuttle boosters were recovered by similar means. The progressive development of Falcon 5 envisaged an advancing evolution in which nine Merlin motors would power the first stage, with a lift-off thrust of 1,250,000lb (5,560kN), about the thrust of early Saturn I rockets of the 1960s, with a second stage powered by a single Merlin. In this configuration Falcon 9 would be capable of delivering 23,000lb (10,430kg) to a low Earth orbit. Falcon 5 would never fly, but Falcon 9 would be the game-changer that very few believed could work.

ABOVE • SpaceX based its rocket development and test facilities at McGregor, Texas. (SpaceX)

BELOW • Hawthorne, California, hosts the SpaceX factory and main headquarters for research and construction. (Steve Jurvetson)

WINGLESS DRAGONS

Cargo delivery by contract

ABOVE • NASA begins the Constellation programme to send humans back to the Moon and plans for a series of commercial contracts to deliver cargo and crew to Earth orbit. (NASA)

BELOW LEFT • Michael Griffin (right) is sworn in as NASA Administrator in 2005, taking the oath from Vice President Dick Cheney to preside over seismic changes as the agency transitions to commercial support for the International Space Station. (The White House)

FAR RIGHT • The first Falcon 9v1.0 series had an arrangement of Merlin engines in a square configuration of three rows of three. (SpaceX)

The real boost to commercial opportunities for private companies came as NASA was beginning to assemble the International Space Station, a major undertaking involving elements built in the US, Russia, Europe, Canada, and Japan. Assembly flights began in November 1998 with the launch of Russia's Zarya module, followed over the next 11 years by a series of Shuttle flights and additional Russian modules until the 500-tonne station had been fully assembled. In all, it took more than 40 flights to build the facility, permanently manned since 2000, with a continuing succession of logistical and resupply missions as well as the delivery and return of astronauts and cosmonauts.

Reacting to the loss of Shuttle *Columbia* on February 1, 2003, it was decided to retire the Shuttle when assembly was complete and develop a commercial programme to provide rockets and spacecraft to do all the logistical and crew delivery work. A seismic change in the way it did business, NASA was offloading the trucking role and appealing to industry, including new start-up companies, to provide these services. It would take time to develop those capabilities as there was no existing programme underway in the United States to do that, and with a desire to retire the Shuttle by 2010 it was imperative to support commercial developments with government subsidies.

The first efforts on that operating model was the Alternate Access to Station (AAS), work which predated the loss of *Columbia* and in 2000 provided some money to several small companies to propose ways in which that could be implemented. A lot of new thinking took place, covering all areas such as managing the risk in opening flight operations involving astronauts carried aboard non-government vehicles and in planning the necessary steps which would have to first demonstrate not only successful flights to the ISS, but the levels of intervention required to ensure safe operations. But it was the loss of *Columbia* in 2003 that proved decisive in moving quickly to get commercial programmes off the ground.

To some extent, the Ansari X Prize of October 4, 2006, which put the first humans into space flying a privately owned rocket ship, validated a fast-track approach to getting NASA offloaded from the financial burden of supplying the space station with cargo and crew. Congress endorsed this approach in 2005 by directing NASA to "develop a commercialization plan to support the human missions to the Moon and Mars, [and] to support low Earth orbit activities." To achieve that an industry day was held at the Johnson Space Center on April 25, 2005, outlining how, uniquely, NASA wanted to procure a service and not a spacecraft and recruiting interest from those companies likely to compete for work.

It would go under the name of Commercial Crew and Cargo Program, or C3PO after the famous robot in the *Star Wars* movies, allowing private industry to own the hardware and lease its services to NASA under contracts. The government would have to support the private effort until those services began, stimulating the commercial sector and building a commercial base for an increasing range of activities outside the specifics of the space station. In February 2006, C3PO became the Commercial Orbital Space Transportation Services (COTS) which would manage and help fund development of the spacecraft, followed by Commercial Resupply Services (CRS) contracts which bind the recipients to delivery services.

Since November 2001, NASA had three administrators running the agency in turn but on April 14, 2005, it received Michael D Griffin who would remain at the helm for almost four years and guide the commercial

ABOVE • *A static fire test of Falcon 9 with a Dragon cargo module on top. (SpaceX)*

programme through its formative period. He would speak the language of New Space advocates, declaring that for him: "the single overarching goal of human space flight is the human settlement of the Solar System, and eventually beyond. I can think of no lesser purpose sufficient to justify the difficulty of the enterprise and no greater purpose is possible."

LEFT • *The end of an era as* Atlantis *comes home on July 21, 2011, completing 135 Shuttle missions since April 12, 1981. (NASA)*

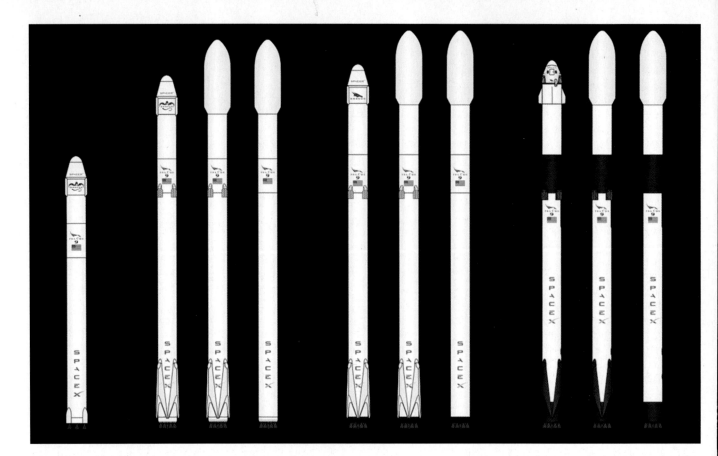

ABOVE • *Falcon 9 was developed into several increasingly more capable variants with optional payload capabilities. (Avialuh and Lubicon)*

BELOW • *The revised configuration of the Falcon 9v1.1 showing the eight Merlin engines arranged in a circle with one at the centre and retractable landing legs within fairings. (SpaceX)*

A founding member of the Mars Society run by Robert Zubrin, co-presenter 14 years earlier of a paper presented to then NASA administrator Daniel Goldin while he was NASA associate administrator for exploration, Griffin had shared a presentation with Elon Musk at the fourth annual Mars Society meeting in 2001, where the idea of growing plants on Mars originated. There have been few times when a NASA administrator was so personally committed to such a radical concept, sharing the dream of commercial enterprise with entrepreneurs who could make it happen.

Critical to achieving long-term goals had been the Vision for Space Exploration presented by President George W Bush on January 14, 2004, in response to the *Columbia* disaster and as a means of reconnecting the public to grand goals on the space frontier. It set up the Constellation programme using legacy Shuttle hardware to take US astronauts back into deep space again. The architects of Constellation envisaged retiring the Shuttle in 2010, using

commercial companies to send cargo and people to the ISS but removing NASA's investment in the station by 2017 and putting astronauts back on the Moon by 2020.

Money saved from Shuttle and ISS would be used to develop the rockets and the spacecraft for returning to the lunar surface and investing in technologies for human expeditions to Mars. An underlying precept nursed by Griffin and his team was that the evolving commercial sector could acquire the technology, the hardware, and the operating experience to continue support for these more ambitious goals, but not everyone felt that way. A more conservative group emerged who questioned the way that would affect control over future human space flight programmes and the challenges posed by integrating commercial with space activities.

What everyone did agree on was the necessity for NASA oversight on the underpinning technologies used and on the safety aspects of the commercial vehicles. Some old hands saw it as an adaptation of the way NASA and the Soviet space engineers had worked together for the joint docking flight between Apollo and Soyuz spacecraft in June 1975. Considering the Russians as the equivalent of a commercial contractor with proprietary rights over certain aspects of its technology, it had been both possible to satisfy NASA safety criteria and achieve a blending of the two very different cultures. Some said that dealing with US corporations and private companies was no less challenging than dealing with the Russians!

Immediately on being sworn in, Griffin allocated $500m over five years to subsidise the commercial start-ups, asserting that the cargo supplies should be well established before moving across to crew transport. Getting the right people involved was crucial to its success and Bill Gerstenmaier, formerly the ISS programme manager, picked Alan J Lindenmoyer to manage the process, effectively seeing it through the transition between C3PO and COTS. Little more than seven months after the industry day to present the

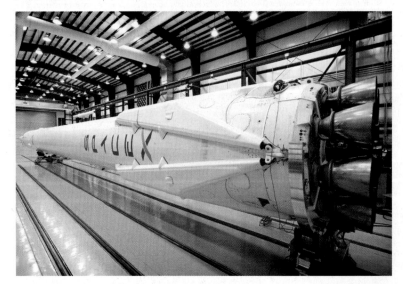

programme, on October 5, 2005, Lindenmoyer chaired the first meeting of the COTS procurement development team, briefing NASA people on what was going to be needed.

Three weeks later, the outline requirements package was finalised explaining what would be required for the cargo transportation hardware through four specifically defined capabilities: an external unpressurised cargo module for delivering payloads to the station and returning with used items to burn up in the atmosphere; an internal pressurised cargo delivery for internal use aboard the habitable modules and carrying a load of 18,520lb (8,400kg); an internal pressurised cargo delivery and intact return module in which experiments and instruments could be returned back down through the atmosphere to a safe landing; and a crew delivery capability with safe landing for returning astronauts.

At this point nobody quite knew when the crew capability would be possible – that seemed a big step forward – but it was recognised that it would require additional funding. Wisely and in parallel with the requirements NASA sought expert advice on how it should interact with the commercial companies and on how to set up the legal framework for smoothing the way to contractual arrangements. It did not want to specify the size and launch frequency, although these factors would be crucial for the companies designing their hardware. In January 2006, NASA had all the documents ready for presenting to the interested parties so that they would have a common framework to work toward in designing their spacecraft.

The demand on cargo to the ISS was very high and would require an estimated 45,750lb (20,750kg) to be uplifted every year. Until it was retired in 2011, much of that was being delivered by the Shuttle or by European and Japanese modules launched by Ariane and H-II rockets, respectively. In addition, the Russians delivered small quantities of air, water, food and supplies via their Progress cargo-tankers, derivatives of the Soyuz spacecraft carrying astronauts and

LEFT • The retracted titanium grid plates on the top of the Falcon 9 first stage which will be used for directional control within the atmosphere during its return to Earth. (SpaceX)

cosmonauts. When the requirement went out for the four stipulated categories on January 18, 2006, it was open to any bidder, so long as the company responding was at least 50% owned by US nationals.

All bids had to be in by March 3, 2006 and 21 companies submitted proposals, some for one or more of the four categories. Big players like Boeing and Lockheed Martin had high hopes of squeezing out mid-size players such as Orbital Sciences Corporation with some largely unknown contenders retaining more hope than substance. NASA wanted two companies under contract to insure that at least one would achieve what they promised, seeing too that a competitive edge would encourage high standards and deliver quickly. In the selection process that ensued, six companies were up for detailed cross-examination but in the final phase on August 15-16, 2006, it was clear that SpaceX had the edge by some considerable margin.

SpaceX was selected on August 18 but of the two contenders for the second contract, NASA chose Rocketplane Kistler (RpK) over SpaceDev, the only proposal

BELOW • A Falcon 9 first stage with extended grid plates at the top after landing on the drone-ship Of Course I Still Love You. *(SpaceX)*

from Russia, with a solid propellant upper stage to deliver the Cygnus cargo module to the space station. SpaceX would develop Falcon 9 and the Merlin engines to lift its Dragon capsule to the ISS. Under the contracts announced on December 23, 2008, SpaceX would fly 12 missions and Orbital would provide eight.

Enter Falcon 9

Development of Elon Musk's Falcon 9 launch vehicle was based on intuitive and innovative design choices using advanced technology and proven techniques. With nine Merlin 1C engines, each delivering a thrust of about 101,000lb (450kN), the first stage would burn for two minutes 50sec at which point it would shut down and separate from the second stage. This was powered by a single Merlin 1C specifically tuned for vacuum thrust, delivering 100,000lb (444.8kN) for five minutes 45sec, placing a maximum payload of 20,000lb (9,072kg) in low Earth orbit.

The design of each stage was conservative, and both would carry liquid oxygen and kerosene propellants. Tank walls and hemispherical end domes were fabricated from aluminium-lithium alloy put together as cylindrical sections connected through friction-stir welding, a technique pioneered on the external tank for the Shuttle but now in frequent use as the strongest known welding available. The upper stage followed the same manufacturing techniques with a design identical to that of the first stage but shorter in length. Throughout, SpaceX used redundancy to increase reliability, with dual igniters in the rocket motors and back-up flight computers to manage stage operations.

Flights to the ISS would have to begin at Cape Canaveral to match the inclination of the station's orbit without detracting from the performance of Falcon 9. Space Launch Complex 40 (SLC-40) was selected and the facility leased to SpaceX from April 25, 2007. Previously, the pad had been used to fly giant Titan launch vehicles, the first taking place on June 18, 1965, but the last of those had been sent up on April 30, 2005, when a Titan IV-B had carried a reconnaissance satellite into orbit. Hardware for the first Falcon 9 flight arrived at the Cape toward the end of 2008 and was erected on January 10, 2009.

ABOVE • A Merlin 1C nears completion, each Falcon rocket is powered by nine engines. (SpaceX)

BELOW • Dragon spacecraft in assembly at the fabrication facility at Hawthorne, California. (SpaceX)

to reach the final stages with a winged spaceplane called Dream Chaser which was designed for flight on an Atlas launch vehicle. RpK had been around for 10 years hoping to raise sufficient money to develop a reusable rocket without attracting support and its failure to meet milestones under the COTS contract caused NASA to cancel that in October 2007. A second competitive round opened later that month, receiving 13 proposals by the cut-off date of November 22.

After a long and protracted series of technical and legal examinations and challenges, on February 19, 2008, NASA selected Orbital Sciences Corporation to be the second COTS winner to parallel SpaceX, presenting its Antares rocket which would be launched from the Mid-Atlantic Regional Spaceport near NASA's Wallops Island, Virginia, facility. Antares was powered by two NK-33 rocket motors

SpaceX spent about $300m developing Falcon 9v1.0 and this was partly subsidised by the COTS contract. A lot was riding on it as preparations began on June 4, 2010, delayed by a few technical problems for approximately three hours. Lift-off occurred at 2.45pm local time and the nine Merlin 1C engines thundered into life, the 157ft (47.8m) tall rocket climbing slowly away from SLC-40. Attached to the top of the second stage was a Dragon qualification module, which would remain attached to it. Pursuing the concept of recovery and reuse, the first stage was intended to deploy parachutes and splash down, but it burned up before the time came to deploy them. Otherwise, the flight had been near-perfect.

A second flight on December 8 was attempted, this being a Dragon demonstration mission with a fully flight-rated capsule. In a 'can-do' spirit that would have shocked early NASA rocket scientists, a crack in the lower part of the second stage Merlin engine's exhaust nozzle was solved by slicing off the lower part of the cone and flying it in that condition. It was another success in delivering the capsule to orbit but the first stage broke up on re-entry before it could demonstrate a deceleration on parachutes. The Dragon capsule was subject to three hours of manoeuvring before it re-entered, splashing down in the ocean as planned.

Although behind schedule by two years, progress in demonstrating technical and operating success was greater than expected and SpaceX received permission from NASA to combine the objectives for the next two flights into one. Instead of flying one mission to approach the space station to within six miles (10km) before returning, followed by a second mission to rendezvous with the station and demonstrate a docking, the third Falcon 9 would go all the way to a docking on the proviso that a series of intermediate steps went according to plan.

Launched on May 22, 2012, it passed all intermediate checks and proceeded through the several steps, each time receiving a 'go' to move to the next until it finally docked with the ISS on May 25, captured by the station's remote manipulator arm controlled by station astronaut Don Pettit with the comment "Looks like we got us a Dragon by the tail!" Cargo Dragons could not dock autonomously and required the manipulator arm to pull them in for a docking. The hatch was opened the following day and station crew floated inside to retrieve some of the cargo it carried, which included 672lb (305kg) of food and clothing, 269lb (122kg) of cargo bags, 44lb (20kg) of science equipment and 22lb (10kg) of computer equipment, a total load of 1,007lb (457kg) uplifted to the station.

Less than six days later the Dragon capsule separated and returned to Earth, clearing the way for future commercial services to begin, the first of which (CRS-1) took place on October 8, 2012, with a docking two days later for the second successful mission to the station, ending when Dragon splashed down on October 28. The second contracted logistics delivery mission (CRS-2) got off the pad at SLC-40 on March 1, 2013, carrying 1,493lb (677kg) of cargo and returning to Earth on March 26 with 3,020lb (1,370kg) of scientific experiments and equipment, the capsule being recovered from the Pacific Ocean a few hours after splashdown.

These five Falcon 9v1.0 flights preceded introduction of an upgraded and significantly more capable launcher, Falcon 9v1.1. This had a reconfiguration of the Merlin engines at the base of the first stage from three rows of three to a ring of eight around the outer periphery of the stage and one in the centre. Designated Merlin 1D, the new motors had a thrust of 147,000lb (653.8kN) producing a total stage thrust of 1,323,000lb (5,884kN) and could be

throttled between 70% and 100% with a nominal operating time of three minutes. The second stage was powered by a single Merlin 1D with a vacuum thrust of 161,000lb (716kN) operating for a burn duration of six minutes 15sec.

With a 30% increase in payload capacity, Falcon 9v1.1 could place 28,990lb (13,150kg) into orbit but that would be when the first stage consumed propellant to depletion. Musk wanted to recover these stages for further flights, which would need some propellant remaining for powering the stage back toward the recovery site and then performing a braking manoeuvre for a vertical soft landing, in which case the payload capacity would be 24,000lb (10,886kg). Engine tests began in April 2013 and a full duration firing was completed two months later. SpaceX was already booking commercial flights for private owners and operators, attracted by the low prices achieved through efficient manufacturing, test, and launch operations. But Musk knew that only by fully recovering each stage could prices really start to tumble and Falcon 9v1.1 was designed to begin that process.

The programme's value to the International Space Station was immense. Dragon was the only cargo module capable of returning to Earth intact with significant quantities of equipment and supplies offloaded from the station and brought safely back to Earth. Europe had its own cargo

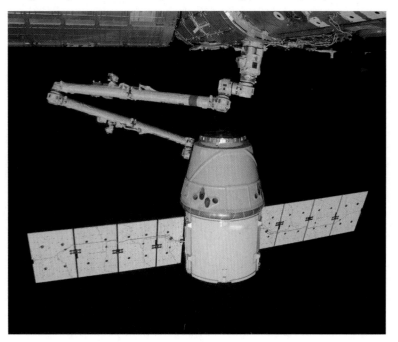

mode, the Automated Transfer Vehicle, built by the European Space Agency as part of its partner deal, but that could not return safely through the atmosphere. Neither could Japan's Transfer Vehicle launched on its H-II rocket, which would similarly be destroyed in the atmosphere after departing the station. Both were developed by industry on government contracts, and none were privately owned.

Bigger and Better

The first flight of the Falcon 9v1.1 took place on September 29, 2013, the first Falcon also from Vandenberg Air Force Base, California, carrying a science satellite into an elliptical, near-polar orbit for the Canadians. The first stage conducted a retro-burn to demonstrate a decelerated landing in the ocean, but it began rolling which starved the engine of fuel causing it to slam into the water uncontrollably. The second flight of this upgraded rocket took place on December 3, 2013, another commercial flight for a US communications company and the first time the second stage motor had fired twice to achieve a highly elliptical trajectory.

And so it would continue, with the launch of a communications satellite for Thailand on January 6, 2014, and the third station resupply flight on April 18, during which it again tried to demonstrate a simulated recovery at sea. For the first time, the first stage had deployable landing legs which on a future flight it would use to land on a platform at sea. This time the simulation worked as it decelerated and slowly touched down on the water, before falling over and breaking up as expected. But it had demonstrated that a safe landing *could* be achieved had it been targeted to a floating platform.

That year, SpaceX launched a further four flights, including three for commercial companies and the fourth resupply mission to the ISS on September 21. It would be followed by seven launches in 2015 but the penultimate flight on June 28 resulted in the first loss of a Falcon 9 when an over-pressurisation in the second stage caused an explosion. It also resulted in the loss of the Dragon cargo module because the parachute recovery system was not wired to activate in such an occurrence.

But high success was achieved on the following flight launched on December 22 when the first stage returned to Earth and touched down on the specially prepared pad at Cape Canaveral named Landing Zone 1, another historic first for SpaceX. That booster (B1019) is on display at the SpaceX headquarters at Hawthorne, California, A second pad, Landing Zone 2 would become available for flights back to Cape Canaveral beginning in 2018.

That initial touchdown on land for safe recovery of a launch vehicle's first stage was historic in itself and another credible first for SpaceX. It was also the initial flight for the next generation of this rocket, the Falcon 9v1.2, sometimes also referred to as the Full Thrust variant, specifically designed for a full recovery and landing from a geosynchronous mission on to a drone ship, for which the Falcon 9v1.1 had been tested as precursor to the fully operational system. With so many orders for launches attracting customers who previously would have flown on a ULA launcher or an Ariane V, SpaceX wanted to combine all the optimised flight capabilities of the geosynchronous path which was the dominant trajectory for the lucrative broadcast and satellite TV market.

Musk pushed hard to introduce some truly creative ways to 'bend' the science and adapt the engineering to squeeze out maximum potential from the existing design. Propellants for Falcon 9v1.2 would be sub-cooled, chilled down lower than they need to be for effective combustion, so as to densify the liquids and obtain just that little bit of extra performance from them. The second stage tanks would be longer, and the first stage would have engineered changes to improve reliability, reduce weight, adjust the layout of the first stage Merlin 1D motors and increase their output, raising lift-off thrust to 1,710,000lb (7,607kN) for a burn duration of two minutes 42sec. The second stage would have a single Merlin 1D delivering a thrust of 188,000lb (836.2kN) for six minutes 37sec.

Falcon 9v1.2 was developed specifically for the commercial launch vehicle market and could place a maximum 50,300lb (22,800kg) in low Earth orbit or send 18,300lb (8,300kg) to a geosynchronous transfer path, the most sought-after slot where most of the revenue-earning customers wanted to place their satellites. It was the market where Europe's Ariane 5, capable of delivering a mass of 15,320lb (6,950kg) to that orbit, got most of its customers. Concurrent with developing a high-performance satellite delivery system, SpaceX had gone far beyond what was required for the cargo contract with NASA and began to progressively build confidence among users and customers looking to reduce the cost of flying their profitable satellites.

Throughout this period, consumer demand for TV and telecommunications services increased exponentially and SpaceX sought to dominate that launcher market by continuing to expand Falcon 9 profitability, lower prices for customers and make the most reliable rocket ever operated. Over time it achieved that and along the way introduced truly revolutionary technology and procedures. One of which was to replace the traditional and complex array of ground radar, tracking systems, computerised ground equipment in protected bunkers and in the number of associated personnel at Cape Canaveral, for a completely new and autonomous flight control and safety system.

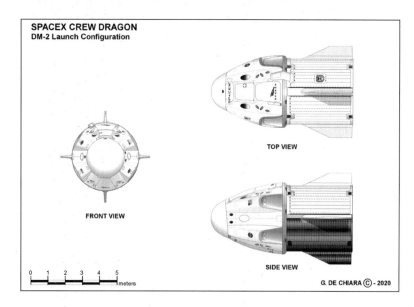

SPACEX CREW DRAGON
DM-2 Launch Configuration

FRONT VIEW

TOP VIEW

SIDE VIEW

G. DE CHIARA © - 2020

This was a major transformation and not only eliminated for its flight operations virtually all the costly and time-consuming network of ground support equipment for pad operations at the Cape. Being a private launch provider requires all services to be hired from the government, either from NASA or the US Air Force, and eliminating that charge dramatically impacts the cost of conducting flight operations and the number of people actively involved in the launch process, from the pad all the way into orbit. Nobody else had done that, a reflection of the installed acceptability of high costs for launch activities. But there was more to come.

In August 2017, SpaceX began flying its Block 4 Falcon 9v.1.2, which retrospectively assigned the three evolutions of the rocket as Blocks 1-3 (Falcon 9v1.0, v1.1 and v1.2). Block 4 incorporated a range of minor improvements and served as precursor to a Block 5 variant which carried that process further. The underlying requirement was in support of a bid to carry astronauts to the International Space Station as an extension of the commercial programme started by NASA when it decided to retire the Shuttle. Plans for that were delayed and the agency stipulated that people would only fly on a commercial rocket that had conducted at least seven flights of a specified launch vehicle without major modifications.

SpaceX started flying Block 5s from May 11, 2018, with a geosynchronous communications satellite sent into orbit for Bangladesh, its seventh launch completed on November 15 clearing the way for flights with its crew-carrying capsule, Dragon 2. Musk also had his eye on obtaining contracts to launch highly classified military payloads into orbit which came with equally constrictive requirements for performance, safety, and reliability. Standardisation

ABOVE • The Dragon 2 vehicle was available in both cargo and crew-carrying variants with each having autonomous rendezvous and docking capability. (Giuseppe De Chiara)

LEFT • NASA proudly boasts its commercial crew programme. (NASA)

BELOW • For several years NASA was unable to get necessary funding for its commercial cargo programme, resulting in delays and added costs. (NASA/Philip McAlister)

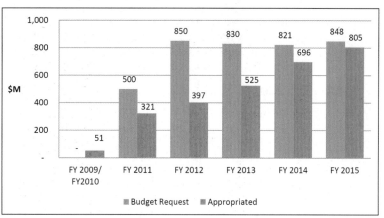

was key to economic operations and SpaceX flew the last Block 4 on June 29, 2018 since when all flights have been with the Block 5.

By the end of that decade, against the predictions of many naysayers, penetration of the global satellite launch market had been impressive, total flights with Falcon 9 for both NASA contracted and fully commercial flights increasing from two in 2010 and two in 2012, three in 2013, six in 2014, seven in 2015, eight in 2016, 18 in 2017, 21 in 2018 and 13 in 2019. The upward trajectory of annual launches for Falcon 9 increased, with 26 in 2020, 31 in 2021, 60 in 2022 and 91 in 2023. If present trends continue, SpaceX will launch more than 100 Falcon 9s in 2024. The majority of these are for commercial customers, SpaceX accounting for 87% of all the mass launched around the world in 2023.

By May 2024, SpaceX had recovered a staggering 303 Falcon 9 first-stage boosters for relaunch, one (B1060) being re-flown 19 times. Moreover, with the exception of the 32 Saturn-series flown in the 1960-70s for NASA missions without a single failure, Falcon 9 has now flown more than 340 times with a 99.4% reliability making it the most successful production launcher in history. Ariane 5 demonstrated a 95.7% success rate compared to China's Long March with 95.3%, 95.1% for Russia's Soyuz launcher (not to be confused with the spacecraft of that name), and Russia's large Proton rocket at 88.7%.

Crew Dragon 2

After the decision had been made in 2004 to retire the Shuttle by 2010 (it actually achieved that in 2011), NASA was keen to get commercial contracts to fly astronauts back and forth to the International Space Station. Europe, Japan, and Canada were able to have their astronauts flown to the ISS aboard the Shuttle as reimbursement for having built hardware for the station at their own expense. The other partner, Russia used its own Soyuz crew-carrier and supplemented small quantities of freight through its Progress cargo-tankers which had been doing those tasks since the Soviet era supporting Russian Salyut and Mir stations.

Until it could replace the Shuttle with a new US people-carrier, NASA would 'buy' seats aboard the Soyuz but at a cost. For that reason, it wanted to accelerate the development of a privately developed spacecraft in much the same process as the COTS programme for cargo delivery to the ISS. The Commercial Crew Development (CCDev) effort got under way on August 10, 2008, with a request for potential contenders to express interest, 36 companies responding by the September

22 deadline of which 18 made it through the initial screening process with eight as finalists.

On December 8, 2009, five companies were chosen for funded agreements, reduced to four in April 2011 and three in August 2012. The two front runners were SpaceX who received $440m for essentially a development of the cargo module technology while Boeing got $460m for its CST-100 Starliner spacecraft. Sierra Nevada got $212.5m for Dream Chaser, the only winged contender patterned after the lifting-body programme and NASA's HL-20. It was expected that the first crew taxi services would be up and running by 2017.

The programme had already been significantly delayed through a brutal and sustained tug-of-war between politicians concerned at using taxpayer money to fund private corporations and those who saw it as merely the latest in NASA's traditional role of stimulating industry before getting out and leaving it to a shared investment. In its annual budget requests in fiscal years 2011-2015, the US government on behalf of NASA received $2.744bn from Congress for the commercial crew programme, $1,105bn less than the amount requested over those five years. That, and delaying tactics from politicians unconvinced as to the value of the programme, cost far more in extended payments and dollars to Russia.

Through a series of supplemental payments an incremental checks on performance, on September 16, 2014, two contenders were awarded definitive development contracts, SpaceX receiving $2.6bn and Boeing getting $4.2bn. Sierra Nevada had already declared a delay in the preparation of its programme and there had been minimal progress. But opinion on the Dream Chaser concept was divided and Sierra Nevada issued a formal complaint which resulted in confirmation of the decision, despite some accusations that the points awarded by an evaluation board had been amended by a senior NASA official to skew the results. But Dream Chaser was not done and would be back.

During development towards its people-carrying goal, Dragon 2 evolved along two separate but parallel branches: Crew Dragon for the astronaut taxi role; and the unmanned Cargo Dragon, taking on improvements developed for Crew Dragon and with a more advanced operating procedure including an autonomous docking capability. SpaceX got the contract to develop Crew Dragon on September 16, 2014, from which would come improvements applied to Cargo Dragon. At first, SpaceX expected to build a new vehicle for every crew flight but that changed, with NASA approval, for repeated launches with recovered spacecraft.

Both versions have a re-entry module and a logistics trunk, the latter not being recovered and ejected to burn

up on re-entry. Cargo Dragon can uplift 7,291lb (3,307kg) and Crew Dragon can carry four people to the ISS, although it has a design capacity for seven crew members. Four SuperDraco rockets are clustered in four groups of redundant pairs, each delivering a thrust of 16,400lb (72.94kN) for aborts which would propel the spacecraft away from a malfunctioning launcher to a distance of more than 2,500ft (762m). Each cluster has two Draco thrusters for attitude control and rendezvous manoeuvres, each of the 16 motors delivering a thrust of 90lb (400N).

Overall, Crew Dragon has a height of 26.7ft (8.1m), a diameter of 13.1ft (4m) and an internal volume of 328ft³ (9.3m³) with a trunk volume of 1,300ft³ (37m³). At launch, Crew Dragon has a maximum weight of 13,228lb (6,000kg) and a return download mass of 6,614lb (3,000kg). Inside, the spacecraft is relatively spacious, aesthetically tuned to the new century in which it would excel and very different to the previously flown ballistic capsules epitomised by the Apollo era. This was no 'flying machine' in the strict sense that it had no wings and could not be flown to a landing pattern in the way the Shuttle could, but it was an efficient and effective answer to the challenge of replacing Russian couches with ones 'made in the USA'.

Control and the monitoring of spacecraft systems was provided through a display of three screens with different programs and multiple presentation options, replacing the several hundred switches, dials and ribbon displays in a typical Apollo spacecraft. With very little lift during descent, this was the first spacecraft of its kind since the last Apollo mission in 1975 but it provided the crew with a broader spectrum of information and greater flexibility in terms of what it could manage in spacecraft systems.

Defined as the Demo-1 mission the first flight of a Crew Dragon (C204) occurred on March 2, 2019, when the unmanned spacecraft was launched on a Falcon 9 to dock with the International Space Station a day later. It returned to Earth after five days docked to the station, having tested essential elements of the vehicle, and demonstrated its operational capabilities. Named *Endeavour*, the first manned Crew Dragon was launched on August 2, 2020, carrying Doug Hurley and Bob Behnken, the first launch of astronauts aboard a US spacecraft in more than nine years and the world's first commercial crew vehicle to carry astronauts into orbit.

The flight was an outstanding success, lasting 63 days during which the astronauts prepared for the Crew 1 visit. That mission was launched on November 16, 2020, the first fully operational, contracted crew delivery flight to the ISS carrying four astronauts, remaining docked to the ISS until May 2, 2021. During that time, on April 5 the spacecraft was relocated to another berthing port on the ISS, demonstrating an additional requirement which other Crew Dragon spacecraft would perform on later missions.

It did not stop there. In the middle of the night on September 16, 2021, SpaceX launched Crew Dragon *Resilience* on an orbital mission carrying four people, the first to fly only private citizens. Named Inspiration4, the mission raised publicity and funds for the St Jude Children's Research Hospital, a charity institution in Memphis, Tennessee. Founded by wealthy entertainer Danny Thomas, although costing $1.7m per day to run, it charges nothing for its services and has grown to include eight affiliated hospitals across the United States. For three days Jared Isaacman, Sian Proctor, Hayley Arceneaux and Christopher Sembrosko orbited the Earth, raising money and support for a hospital whose founder touched the hearts of many when he declared that "No child should die in the dawn of life."

LEFT • Part-time astronauts (from left), Sian Proctor, Jared Isaacman, Christopher Sembroski, and Hayley Arceneaux, crew members for the Inspiration4 mission raising funds for a children's hospital. (NASA)

BELOW LEFT • Crew Dragon Resilience *is recovered from splashdown in May 2022 after 167 days in space. (NASA)*

BELOW RIGHT • Crew Dragon docks automatically to the International Space Station with a hinged nose cone exposing the docking collar and capture latches. (NASA)

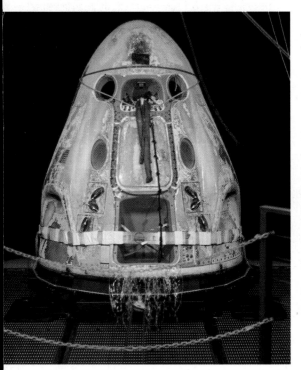

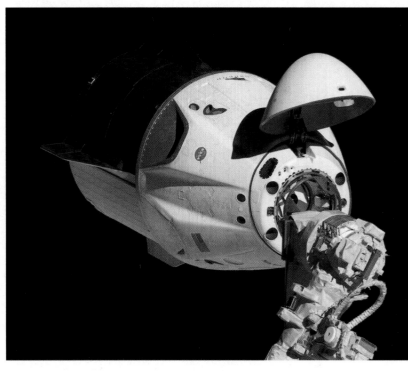

FELLOW TRAVELLERS

Catch-up companies join the club

Dominating the commercial crew and cargo programmes, SpaceX was only one of several different US organisations receiving NASA contracts for developing competitive services. The space agency wanted a back-up for each of the different functions it placed contracts for so that any problem with either supplier would not halt the delivery of logistics and astronauts to the International Space Station.

There was good reason for this. Previously, with only one type of vehicle lifting both crew and cargo to the station, the loss of *Challenger* in 1986 and *Columbia* in 2003 had grounded the Shuttle for a total of three years and two months. In those intervals, only Russia could use frequent flights of its Progress cargo-tankers for station replenishment and Soyuz spacecraft for carrying crewmembers up and down to space. It had been a mistake to combine both functions into the Shuttle.

Under the commercial programme, NASA wanted a second supply chain to guarantee continued delivery and sustainable operations on the ISS.

In addition to the smaller Russian spacecraft, five Automated Transfer Vehicles (ATVs) from the European Space Agency were launched to the ISS by Ariane 5 between 2008 and 2015, each supplying up to 16,900lb (7,665kg) of cargo, prior to the introduction of the US commercial services, which were dominated by Dragon since 2010 before it was retired in March 2020 and by Cargo Dragon from December 2020 to the present.

The initial COTS programme that saw SpaceX introduce its Dragon cargo-carrier also gave Orbital Sciences funds for its own proposal, the Cygnus spacecraft, less capable than its competitor and unable to either dock to the ISS or bring freight back through the atmosphere to Earth. Under the Commercial Resupply Services (CRS) deal, on December 23, 2008, Orbital

BELOW • Orbital Sciences chose the Mid-Atlantic Regional Spaceport off the Virginia coast as the launch site for its Antares rockets carrying Cygnus cargo modules (NASA)

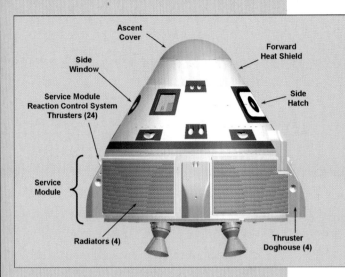

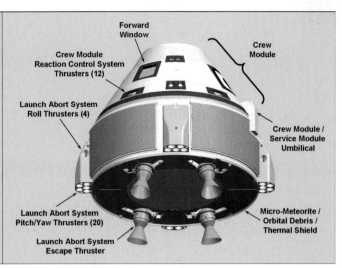

Side Window

Ascent Cover

Forward Heat Shield

Side Hatch

Service Module Reaction Control System Thrusters (24)

Service Module

Radiators (4)

Thruster Doghouse (4)

Forward Window

Crew Module Reaction Control System Thrusters (12)

Crew Module

Launch Abort System Roll Thrusters (4)

Crew Module / Service Module Umbilical

Launch Abort System Pitch/Yaw Thrusters (20)

Launch Abort System Escape Thruster

Micro-Meteorite / Orbital Debris / Thermal Shield

Sciences received a $1.9bn contract to deliver eight cargo delivery flights to the station by 2016. As defined under the four options laid out by the COTS plan, like Dragon, Cygnus would have an unpressurised trunk and a pressurised module, the payload being 4,410lb (2,000kg) and the complete assembly weighing 14,553lb (6,600kg) at launch. With a total length of 16.7ft (5.1m) and a diameter of 10ft (3.07m), Cygnus produced 3.5kW of electrical power from two solar panels on opposing sides of the trunk.

Under the COTS agreement, Orbital Sciences received $171m to seed development with the company matching that with its own investment. Like SpaceX, Orbital Sciences would own the hardware and lease its use to NASA. The

company had little experience with large rockets and decided on a relatively conventional, off-the-shelf series of stages and rocket motors for the launcher to send Cygnus on its way. It would be carried into orbit by an Antares rocket consisting of a liquid propellant first stage and a solid propellant upper stage.

The first stage was contracted to a company in the Ukraine with extensive experience building large ballistic missiles and would be powered by two Aerojet AJ26 engines, adapted from the powerful Russian NK-33 deigned by Nikolay Kuznetsov during the Soviet era. They would provide a lift-off thrust of 734,000lb (3,264kN) for three minutes 55sec. The second stage comprised a Castor 30-series solid propellant rocket motor, the specific type of which was tailored to evolving generations of launcher, producing thrust levels of 58,200lb (258.8kN) to 107,000lb (475.9kN) for different variants.

The first launch occurred at the Mid-Atlantic Regional Spaceport near NASA's Wallops Island, Virginia, on April 21, 2013, which was more a demonstration of the Antares rocket than the payload, comprising a mass simulator of the Cygnus spacecraft. The second launch on September 18 carried the first flight-rated Cygnus loaded with 2,780lb (1,260kg) of provisions and it docked with the ISS on

ABOVE • Essential elements of CST-100 Starliner with the disposition of thrusters and equipment around the exterior of the spacecraft. (Boeing)

LEFT • The rear end of the Antares rocket displaying the two NK-33 rocket motors which powered the early series. (Orbital Sciences)

LEFT • Much later than the SpaceX Dragon, the Cygnus cargo module is seen approaching the International Space Station as envisaged by an artist. (Orbital Sciences)

ABOVE • *The Enhanced Cygnus uses Orbital ATK Ultraflex solar arrays replacing the wing-mounted cells on the earlier module. (NASA)*

September 29, berthed by the station's remote manipulator arm. It remained docked for more than 22 days and was de-orbited to destruction in the atmosphere on October 23. The first officially contracted flight occurred on January 9, 2014 and was followed by the second on July 13.

Although well behind the original plan which envisaged cargo delivery flights starting in 2010, Cygnus was a simple but highly efficient carrier, albeit with limited capabilities in that it could not autonomously dock with the station or bring payloads back from the station. Nevertheless, the programme provided a valued supplement to cargo delivery requirements until the Antares rocket failed a few seconds after lift-off on October 28, 2014, while starting its third contracted delivery flight. The spectacular detonation of the powerful rocket was all the more dramatic as it floodlit the nocturnal scene in a colourful fireball.

Orbital Sciences had a contract to deliver cargo and with Antares out of action Cygnus would fly on an Atlas V until a redesigned and more capable Antares 230 rocket became available, two launches occurring on December 6, 2015 and March 23, 2016. With first stage engines replaced

by more powerful ex-Soviet RD-181s, the initial flight of Antares 230 took place on October 17, 2016, carrying Cygnus to a successful docking with the ISS six days later but a third Atlas launch was necessary, on April 18, 2017, before Antares was fully up to schedule.

By this time Orbital Sciences had recovered and the new launcher was operational from the next cargo flight on November 12, 2017, carrying an Enhanced Cygnus with modifications and greater payload capability which had first flown on one of the preceding Atlas V flights. The Enhanced Cygnus had a total length of 21ft (6.39m) and could carry 8,268lb (3,750kg) in a pressurised volume of 953.5ft^3 (27m^3) versus 667ft^3 (18.9m^3) for the standard Cygnus. For both versions, the structure of the cargo module is provided by Thales Alenia, Italy.

Changes to the programme brought about by the political situation over the Russian invasion of Ukraine in February 2022 accompanied a decision to introduce Antares 300, the last of the 230-series rockets launching an Enhanced Cygnus on August 2, 2023. Perhaps ironically, Elon Musk's Falcon 9 was used to launch the next Enhanced Cygnus on January 30, 2024, and will be followed by at least two more this year before Antares 300 is expected to take over from June 2025.

The new Antares launcher will be powered by seven Firefly Miranda rocket motors producing a stage thrust of 1,600,000lb (7,116.8kN) at lift-off. To date, Orbital Sciences, owned by Northrop Grumman since 2018, has flown 21 Cygnus and Enhanced Cygnus spacecraft to the ISS with additional contracts from NASA sustaining the valued role this storied vehicle has performed. In addition, SpaceX has flown 30 Dragon and Cargo Dragon missions and both providers have each suffered just one loss.

Starliner – The Late Runner

From the start of NASA's competitive Commercial Crew Program in 2010, the agency sought to have two commercial contractors to back-up each other in the event of one suffering a catastrophic failure preventing an early return to flight. Boeing pitched its proposal around a reusable spacecraft looking like a slightly larger version of the Apollo Command Module attached to an expendable service module.

RIGHT • *Cargo being placed in the Cygnus module for delivery to the International Space Station. (Orbital Sciences)*

Known as the CST-100 — for Commercial Space Transportation, a vehicle capable of exceeding the 100km Kármán line, generally regarded as the demarcation line between the atmosphere and space — it was named Starliner, capable of carrying a crew of up to seven people to low Earth orbit where it could remain for up to seven months. NASA selected Boeing as one of the two winners in a competition that also gave a development contract to SpaceX, each being required to start providing operational services from 2017.

Unlike Dragon spacecraft, which would be launched by the SpaceX Falcon 9v1.2, Starliner would fly on an Atlas V from SLC-41 at Cape Canaveral. This pad had been active since the first launch from there on December 21, 1965 when it supported a Titan IIIC before flying the Atlas V from August 21, 2002. Atlas V is very different from the original launch vehicle of that name which put the first four Earth-orbiting American astronauts into space during 1962 and 1963, since when it had never carried a manned vehicle until selected for Starliner.

With a total height of 16.5ft (5.03m) and a diameter of 15ft (4.56m), Starliner weighs 29,000lb (13,000kg) fully loaded and has an internal volume of 390ft^3 (11m^3). Boeing adopted a unique, weldless, spun-form structure with a capability to fly up to 10 missions with a six-month turnaround between flights, individual crew modules being assigned to alternating launches. On a standard mission it would fly four crew members to the ISS, the same as SpaceX's Crew Dragon. Starliner uses a unique Boeing heat shield and carries solar cells on the outer face of the service module providing 2.8kW of electrical power. Four RS-88 abort motors, each with a thrust of 39,500lb (176.6kN), are located at the base.

The contract for Starliner came on September 16, 2014, after Boeing had received $460m in development money, with a $4.2bn contract for full scale manufacturing and test including at least one crewed demonstration mission. There had been a plan for Starliner to also compete for the cargo resupply role but that was dropped in September 2015 and a few months later the 2017 flight date began to slip. In December 2015, former NASA astronaut and Shuttle commander Chris Ferguson joined the Starliner programme as head of its crew development office and as one of the two crew members on the first flight but that changed over time and Barry Wilmore was eventually assigned to the first crewed flight to be accompanied by astronaut Suni Williams as the NASA-designated appointee.

As the programme flagged and technical delays, funding shortfalls, and testing troubles plagued Starliner, the programme flight schedule underwent numerous changes. By the time Wilmore and Williams had been assigned, the flight was to last two weeks at the ISS but that changed to six months, much as Dragon had conducted on its first crewed mission. That was again changed to a flight of eight days. Insiders claim that with its stronger familiarity with many NASA programmes, the space agency had declined the scrutiny it applied to newcomer SpaceX, where

it accorded considerable attention to technical details, manufacturing and assembly and flight crew operations.

The first unmanned Orbital Flight Test (OFT-1) was launched on December 20, 2019, for a seven-day docking at the ISS but a series of technical flaws caused the mission to be aborted and Starliner returned to Earth two days later, having not visited the station. The only 'first' achieved by this flight was the touchdown on land for a ballistic US spacecraft returning from orbit, a series of cushion-bags softening the impact. Previous ballistic capsules from the Mercury, Gemini, and Apollo programmes had landed on water.

Post-flight examination of the mission and inspection of the crew module revealed several software problems and the communication equipment also revealed inadequacies. These problems were attributed to Boeing and the company agreed to launch a second mission, OFT-2, at its own expense. But as the investigation progressed, the schedule slipped further still and plans for a launch in August 2021 were dropped when Starliner's propulsion system also displayed problems with its control systems.

Finally, on May 19, 2022, OFT-2 was launched by Atlas V on a successful, seven-day flight to the ISS but without a crew. It carried 800lb (362kg) of cargo divided between NASA and Boeing supplies but there were still a few technical failures, corrected through manual workaround procedures. It returned to Earth on May 26 to a further series of delays brought about by concerns raised as engineers probed every element of the vehicle. First there were problems with the parachutes — it was found that bracing tape presented a fire risk — and then by technicians discovering design flaws in the parachute lines due to incorrect testing. Drop tests were required to prove corrective processes and the date for the first crewed flight moved inexorably to the right yet again.

After a delay of almost two years since OFT-2, the third spacecraft (S3) was finally scheduled for a crewed launch carrying Wilmore and Williams early in the second quarter of 2024 and the first attempt to fly occurred on the late hours of May 6 local time. This time it was the fault of Atlas V to fail the mission when a valve on the upper stage, a Centaur, showed unexpected fluctuations causing cancellation. That brought a delay while the problem was addressed. With the reputation of Boeing already challenged by multiple issues regarding its airliners, nobody wanted another corporate setback, while all personnel at both the company and NASA were uncompromising on risk mitigation and crew safety.

ABOVE • The Boeing Starliner designed to carry crew to and from the ISS, depicted here with conical pressurised crew module and the service module, about to dock. (Boeing)

LEFT • Starliner approaches the International Space Station on its uncrewed OFT-2 flight in May, 2022. (NASA)

CITIZEN
MOONBASE

Giant rockets and lunar missions
by private contract

As noted in previous chapters, SpaceX was not the only private company to own and operate a logistical delivery system for NASA, supplying the space station with food, clothing, equipment, and sundry items along with Orbital Sciences, now owned by Northrop Grumman. To date, however, SpaceX is the only one to operate a routine, commercially-run taxi service ferrying astronauts back and forth to the International Space Station. Although in time, perhaps in 2025, it is likely to be joined by Boeing with its Starliner spacecraft cleared for routine, contracted flights.

At present, NASA plans to deorbit the International Space Station in early 2031 but an expectation that commercial space station operators will fund and develop a series of smaller stations for research to follow the ISS is by no means certain. However, China is developing its own space station and has already built a large orbiting research facility called Tiangong which it routinely occupies for scientific experiments, attracting international companies and governments in cooperative ventures.

At present, government-level partners in the NASA-led ISS (Europe, Canada, Japan, and Russia) have no plan to replace it and Russia has already expressed interest in going it alone or joining with China in its Tiangong station. Neither is the political climate conducive to future partnership deals with Russia. So, what of the commercial opportunities when the ISS is no more and there are no destinations for Earth-orbiting astronauts to visit? Is there a future at all for commercial crew-carriers and if so, to

BELOW • Without the leasing of the Starship Human Landing System from SpaceX, NASA would be unable to get back on the Moon, an event planned for no earlier than September 2026 and likely to be delayed by at least one or two years. (SpaceX)

where? That question may be answered by way of the larger space sector in global economies.

SpaceX has invested heavily in the development of Falcon rockets for commercial markets, organisations and governments around the world sending satellites into space, seizing ownership of a global demand for satellite-launchers. Falcon rockets now launch around 100 times and more each year, carrying the vast majority of the world's satellite traffic. These satellites are for GPS navigation devices, environmental monitoring, climate analysis, telephone calls, internet services, weather forecasting and TV broadcasters, for scientific use, commercial revenue, military activities, and public entertainment in everything from satellite-based TV transmissions to global gaming.

With electronic microminiaturisation and high-power transmitters in tiny packages, direct internet to cell-phones and tablets from low-orbiting satellites is already bringing the digital world to even the most remote and isolated areas on the planet. Which is driving commercial providers such as SpaceX with its Starlink system, calling for 12,000 satellites in low Earth orbit providing direct access to the internet for everyone, everywhere. SpaceX is not alone, OneWeb plans a network of 648 satellites, Project Kuiper will provide Amazon with 3,236 satellites and a commercial operator in Russia expects to send up 640 satellites for a similar purpose. And there are more emerging every year.

An equally expanding market covers both civilian and military needs for detailed optical imaging of large areas of the Earth's surface. Local and national government planners use satellite information for monitoring the changing demographic between rural and urban areas, mapping tree cover, collecting data about green-field and brown-field areas, monitoring coastal erosion, sea-state mapping, changes to ice layers, glacial drift, and the state of national transportation routes, in addition to mapping future roads and railways. Digital and radar imaging is also used for military purposes and for maintaining communications between combat units as well as providing the electronic infrastructure for organising and operating aircraft, warships, submarines, and ground forces on a scale unimaginable even a few years ago.

All this activity has provided lucrative business for SpaceX, incurring headaches for established launch providers such as United Launch Alliance with its Delta and Atlas rockets, Europe with its Ariane series, Japan with its H-II, and subsequent vehicles, as well as Russia and China losing out on the international markets to this New Space disruptor from the United States.

As early as 2003, Elon Musk recognised the need for powerful launch vehicles sending up massive communications satellites to geostationary positions where they appear to remain stationary in space as they make one orbit every 24 hours. That was already a lucrative market. Falcon 9 had not yet flown when Musk recognised the need to grow it still bigger for sending heavier satellites to where the most customers had their revenue-earning relay stations in space. Rather than develop a new generation of powerful rockets for an increasingly large first stage to produce the required thrust, Musk opted to cluster three Falcon 9 first-stage sections together, side-by-side, with a Falcon 9 second stage on top of the centre core unit to produce a launch vehicle of phenomenal power.

It was called simply Falcon Heavy, with a lift-off thrust of 5,100,000lb (22,684kN) and a maximum payload capacity of 141,000lb (63,957kg) to low Earth orbit or 59,000lb (26,762kg) to a geosynchronous transfer path. It could also send 37,000lb (16,783kg) out of Earth orbit and toward

ABOVE LEFT • A Starlink satellite terminal displayed by Steve Jurvetson, typical of equipment used by 2.7m subscribers as of April 2024. (Steve Jurvetson)

ABOVE RIGHT • Elon Musk poses with a model of Falcon Heavy, an assembly of three Falcon 9 core stages strapped together, the outer ones serving as boosters which separate prior to the core stage. (SpaceX)

LEFT • A Falcon 9 carries a communication satellite on the first re-flight of an orbital-class rocket, March 30, 2017, underscoring the commercial return from multiple launches. (SpaceX)

The two side boosters separated at two minutes 34sec and returned to a visually spectacular landing together on adjacent pads at the Kennedy Space Center while the core stage burned for a further 33 seconds but failed to make it down successfully and was lost. Powered by a single Merlin 1D with a vacuum thrust of 210,000lb (934kN), the second stage burned for six minutes 37sec, pushing the stage and its Tesla Roadster payload on a curving path to a distance just beyond the orbit of Marts as planned. The trajectory will continue to circle the Sun, reaching just inside the orbit of Earth at its closest approach and out to deep space again, endlessly repeating that cycle for millions of years.

The second flight of Falcon Heavy carried a large communications satellite, Arabsat 6A into a geosynchronous transfer path on April 11, 2019. With a weight of 14,255lb (6,465kg) it was Falcon Heavy's first commercial flight and this time the two side boosters landed successfully back at the Kennedy Space Center, followed shortly thereafter by the core stage on the landing drone-ship *Of Course I Still Love You*.

By this time, SpaceX had booked sufficient commercial payloads to pay for the cost of developing Falcon Heavy, justifying Musk's original idea to corner the large satellite market. This was followed by the third flight on June 25, 2019, lifting 24 small satellites for the US Department of Defense, prior to a classified payload for this customer on November 1, 2022, with added demands on the second stage which was required to coast for a long time around the Earth before placing part of the 8,270lb (3,750kg) payload directly into geosynchronous transfer orbit.

Flights came under increasing demand, the fifth launched on January 15, 2023, being the second classified mission with the same payload mass as the preceding flight, together with a grey painted band around the second stage to ensure thermal stabilisation during a long orbital coast. The next two flights were for civilian commercial companies, a payload of 13,700lb (6,214kg) going up on

Mars. This would give it the greatest lift capability of any rocket in history, except for the uniquely utilised Saturn V for Apollo Moon landings and Russia's Energia which flew only twice before the collapse of the Soviet Union.

But the precise nature of Falcon Heavy was not immediately apparent, as development of Falcon 9 continued toward its first flight in 2010 and continuous changes affected what was to be the template for the later rocket. Formally presented to the potential customer market in April 2011, just after Dragon docked with the station for the first time in May 2012 Intelsat announced that it had selected Falcon Heavy to launch some of its satellites. It was the first commercial order for this rocket.

A new test stand for Falcon Heavy was built at McGregor, Texas, where engine development had taken place for the standard Falcon 9 variants and where soon tests with the triple stages and their 27 Merlin 1D rocket motors could be fired, those activities starting in May 2017. These tests were completed by September and the first static test of a full Falcon Heavy took place on January 24, 2018, on Launch Complex 39A at NASA's Kennedy Space Center where SpaceX had leased it and had conducted extensive changes from the time it supported Shuttle missions and previous Apollo Moon flights.

By this time expectations were high that SpaceX would pull off another spectacular 'first' but as the launch date neared, Musk himself urged caution and appeared nervous, citing that he believed it only had a 50:50 chance of success and that he would be happy if it didn't fail so catastrophically that it destroyed the launch pad! But he had another surprise up his sleeve. In December 2017 he had tweeted that FH 1 would carry his personal Tesla Roadster and that the trajectory on this engineering test flight would be targeted for the orbit of Mars while playing *Life on Mars* through the space-to-ground communications link.

And so it was on that day when the entire personnel at SpaceX in Hawthorne, California, and at the Kennedy Space Center witnessed the largest rocket of the New Space age thunder away from the launch pad into a Sun-filled afternoon sky on February 6, 2018, to the voice of David Bowie. It was almost immediately followed by a new anthem to the era of commercial flight accompanied by a crowd of seemingly very young SpaceX staffers setting up a ghetto-blaster thundering away with Elton John's *I'm Still Standing* as they danced and partied alongside the Vehicle Assembly Building, witness to the impossible that had just unfolded before their tear-filled eyes. Both tunes became iconic reminders of a quite remarkable day when Musk's Tesla headed for Mars, an emotion only those who have stood in the rocket's red glare can fully understand.

May 1, 2023, followed by the heaviest communication satellite, weighing in at 20,300lb (9,208kg), for EchoStar on July 29, 2023. The eighth Falcon Heavy flight on October 13, 2023, carried NASA's 5,750lb (2,608kg) Psyche spacecraft into a heliocentric orbit around the Sun toward the asteroid of that name.

On December 29, 2023, the ninth Falcon Heavy lifted the winged X-37B spaceplane into a high Earth orbit of undisclosed properties for the US Air Force. Weighing approximately 14,000lb (6,350kg), this was the fourth flight of the third vehicle of this type to fly and the seventh mission overall, flights which previously had been launched on Atlas V rockets operated by United Launch Alliance. Three flights are scheduled for 2024, including NASA's Europa Clipper, a large spacecraft weighing 13,371lb (6,065kg) destined to conduct multiple fly-bys of Jupiter's moon Europa beginning shortly after it arrives in April 2030. Four Falcon Heavy flights are already booked for 2025 and two for 2026.

With the retirement of the Delta IV in early 2024, Falcon Heavy has a clean sweep on the launch of large payloads, for now at least until competitors such as the Vulcan Centaur from United Launch Alliance and New Glenn from Blue Origin come along, for which see pages 110-113. At present, SpaceX quotes launch prices equivalent to $1,065/lb ($2,350/kg) for a launch to low Earth orbit and $2,549/lb ($5,630/kg) to geosynchronous transfer orbit. Equivalent cost quotes for a Delta IV immediately prior to its retirement were $5,597/lb ($12,340/kg) and $11,171/lb ($24,630/kg), respectively. The numbers walk the talk.

Starship to the Moon

Long before he had any success with the Falcon 1 rocket, around 2003 Musk nursed hope of a truly gigantic rocket capable of placing more than 100 tons (220,500lb) in orbit using a launch vehicle which he dubbed the BFR – Big Falcon Rocket, or where the 'F' stands for something else. It didn't stop there, however. While seasoned veterans of

countless rocket programmes poured scorn on this young upstart without any experience or having sent anything into space, Musk went blithely along presenting his ideas to forums, social groups and at conferences.

As his success grew, he gained a wider audience and when his Falcon 9 began to take orders away from long-established launch vehicle companies, many of them adept at marketing their rockets, offering cut-price, bulk-launch deals and providing ancillary benefits to suck in the burgeoning customer base, they began to listen. More so when he showed sustained and reliable return of his core first stage elements to soft landings and multiple re-uses. But they only really started listening when he offered basement-price tickets for satellite operators.

Aside from the commercial business, the real attraction for Musk in building such a colossus was his dream of getting to Mars and the rocket was dubbed the Mars Colonial Transporter but that was later changed to the Interplanetary Transport System because his aspirations

ABOVE • Elon Musk donated his personal Tesla Roadster to ride on the first flight of Falcon Heavy which would carry it deep into the solar system. (SpaceX)

LEFT • The SpaceX engine test bunker at McGregor Rocket Test and Development Facility, Texas, where Falcon engines go through demonstration runs. (Steve Jurvetson)

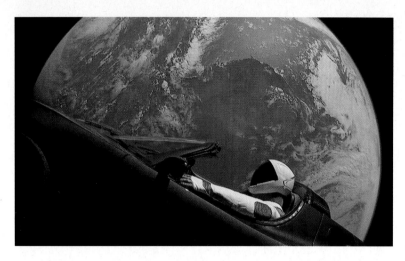

ABOVE • One of the first images beamed down from cameras on the Tesla Roadster attached to the Falcon Heavy upper stage showing Earth in the background. (SpaceX)

BELOW • The effects of the space environment can be seen on the car as the camera looks back at the Starman manikin seated in the Tesla Roadster as they speed away from Earth. (SpaceX)

went beyond the Red Planet, as his marketing graphics would show with imaginative flights around moons of Jupiter. By 2019, the BFR was being marketed under the slightly more acceptable name of Starship, a crew-carrying upper stage to a giant core first stage which quickly became known as the Super Heavy, carrying Starship as the second stage.

Both Super Heavy and Starship were designed to be reusable, what passed for the upper stage being a giant spaceship in its own right with a large void at the forward section capable of carrying more than 100 people. The entire stack would have a height of 398ft (121.2m) and a lift-off mass of 11,000,000lb (4,989 tonnes) with a thrust of 16,700,000lb (74,281kN). Only loosely defined by role, Starship was capable of independent operation and could be used for a wide variety of missions, although Musk saw it as a launcher for several hundred small internet relay satellites such as the SpaceX Starlink system.

Starship and its Super Heavy booster were funded from Musk's own financial resources, a man with a net wealth of $200bn, and that gave him the freedom to think away from the commercial imperatives that had dominated the

success of SpaceX through its cargo and crew-carrying Dragon vehicles and the development of the independent launch system, Falcon 9, which was now an outstanding success as a global launch provider. While professing to be building a vehicle capable of supporting the mass migration of people to Mars, it would be to the Moon that Musk got the first deal for Starship and its first, formal contract – directly from NASA.

When, in 2004 the space agency decided to replace the Shuttle with commercially-contracted support for the International Space Station and return to exploration of deep-space destinations starting with the Moon, the Constellation programme was charged with achieving that. On September 15, 2006, Lockheed Martin got the contract to build Orion, an Apollo lookalike with much the same objective of supporting Earth orbit and circumlunar missions and consisting of a crew compartment and a service module. Adopting the Apollo mission profile, Orion would remain in lunar orbit while a lunar lander named Altair would carry astronauts down to the surface in a vehicle much larger than Apollo's Lunar Module.

The launch vehicles for Constellation included Ares I, a single shuttle, solid rocket booster with an upper stage for flying Orion on Earth-orbiting missions, and Ares V comprising a very large core stage, two solid rocket boosters and a powerful upper Earth Departure Stage (EDS) for placing Altair in orbit to await Orion on an Ares I. Thus docked, EDS would push that stack out of Earth orbit and across to the Moon. As noted previously, while all this was underway, commercial contractors SpaceX, Orbital Sciences and Boeing would support the International Space Station.

Ambitious and expensive, the Constellation programme was costed to be far higher than projected NASA budgets could support and in February 2010 President Obama cancelled it. At the insistence of Congress and in furious response, the following year NASA had a revised plan which replaced the Ares V with a less powerful version known as the Space Launch System (SLS), or the 'Senate Launch

LEFT • When NASA turned to the Constellation programme to carry astronauts back to the Moon it envisaged a very large lunar lander called Altair (left) and the Orion spacecraft before it was cancelled in 2010, Orion being the only survivor. (NASA)

System' as some wags call it, retaining development of Orion for unspecified deep-space missions.

In December 2017 President Trump announced the goal of returning to the Moon using commercial companies to ease the financial burden. By this date NASA had suffered several delays to the SLS and Lockheed Martin had been unable to develop the service module for Orion at an affordable price. In the same way it had received separate modules for the International Space Station in exchange for visits to the station, NASA closed a deal with the European Space Agency for them to build what was now called the European Service Module (ESM) in exchange for seats aboard Orion for ESA astronauts. Neither could the NASA budget afford the lunar lander. Instead of building it in-house, it launched a competition and on April 16, 2021, chose SpaceX to adapt Starship for that job.

The new programme is called Artemis and Moon landings will involve a Starship HLS placed in Earth orbit by the Super Heavy first stage and refuelled with propellant from between 12 and 17 Starship-tankers launched consecutively so that it can power itself to lunar orbit. There, it will await Orion launched by NASA's big SLS rocket, which is less than half as powerful as the SpaceX Super Heavy. In lunar orbit, Orion and Starship-HLS will dock so that astronauts can transfer down to the surface. At the end of several days, Starship-HLS will lift-off and rendezvous with the lunar-orbiting Orion for the return to Earth.

Every crewed lunar mission to the surface involves this number of launches and development of Super Heavy and Starship is critical to getting US astronauts back on the Moon for what NASA anticipates will be a semi-permanent base. From there it will conduct scientific exploration and gather information about the suitability of lunar resources for use back on Earth. If that proves to be so, the commercial sector could grow exponentially, moving mined materials to Earth and supporting a completely new industry. Early studies indicate numerous minerals are there at the surface but the possibility over ownership and extraction has deep political overtones.

China has claimed many times that it intends to put its own astronauts on the lunar surface by 2030, which may be not too far behind the Americans with its Artemis programme. Launched on November 16, 2022, Artemis 1 was an unmanned flight of Orion on the first SLS flying high above the Moon around the far side before looping back and returning to a splashdown in the Pacific Ocean. Carrying the first crew to fly in Orion, Artemis 2 is presently scheduled for September 2025 and will repeat that as a test of the spacecraft's systems before the first lunar landing attempt planned for a year later.

But that requires SpaceX to first show safe and reliable operation with Super Heavy and Starship and NASA requires it to perform a full sequencing of orbital tanking and Moon landing operations uncrewed to demonstrate that it can do all that and safely before landing humans on the Moon. This is now a colossal international effort where the Europeans are building one of the two Orion elements, and the commercial sector is opening the window again on renewed exploration of the lunar surface. The stakes could hardly be higher, a new Space Race is on with more than national pride at stake and the clock is counting down.

BELOW • This graphic displays the complex integration of NASA's Orion spacecraft launched on an SLS and the Starship HLS in the current plan for sending Artemis 3 to the Moon. The number of Starship depot, tanker and HLS flights could require up to 17 separate launches. (NASA)

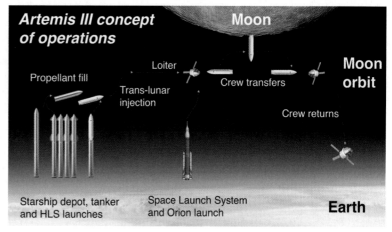

Artemis III concept of operations

Moon

Propellant fill

Loiter

Crew transfers

Moon orbit

Trans-lunar injection

Crew returns

Starship depot, tanker and HLS launches

Space Launch System and Orion launch

Earth

ABOVE • Starbase at Boca Chica, close to a nature reserve and Brownsville, Texas, from where SpaceX launches Super Heavy/Starship rockets. (Jenny Hautmann)

ABOVE • Seen from above, a Super Heavy booster gets a full-thrust test on its launch pad prior to mating with a Starship for flight. (SpaceX)

MUSK CALLING MARS

SpaceX prepares for a mass migration to the Red Planet

O n the evening of Sunday, October 30, 1938, listeners tuned in to the CBS Radio Network were alarmed to hear that a strange object had fallen to Earth near Grover's Mill, New Jersey. Interspersed with advertisements and other news bulletins, the broadcast began to unravel a terrifying conclusion that the United States was being invaded and that the intruders were from Mars. What listeners unknowingly heard was a reading by Orson Welles of H G Wells' book *The War of the Worlds*, in which Martians ravaged the planet before falling prey to a virus against whose deadly effect they had no immunity.

The broadcast was an all too realistic rendition of a science fiction novel published in 1898. Not a lot of people were listening that night when the famous American actor brought a radio rendition of its chilling message – that life may not only exist beyond Earth, but that it is probably hostile and that the nearest of those threats comes from Mars – but it scared many of those who were. That possibility claimed scientific support from the Italian astronomer Giovanni Schiaparelli, who in 1877 had postulated that seasonal changes to the surface of Mars could be agricultural crops planted by intelligent beings.

The reality was that when Welles made his broadcast nobody knew for sure whether alien life existed or not,

with various observations holding the possibility of an Earth-like environment. When the rocket engineer and author Wernher von Braun published a series of articles in *Collier's* magazine during the mid-1950s, after his classic book *The Mars Project* of 1953, it added energy to the debate when it urged an expedition to Mars to discover whether life was present on the surface. That year, in the UK the BBC Light Programme began serialising a science fiction play *Journey Into Space* in which expeditions to Mars were the ultimate target.

With the emerging space programme of the late 1950s, questions about life on Mars lay unanswered until Mariner 4 sent the first pictures back to Earth following a close fly-by on July 14, 1965. Although they could only show features at best two miles (3.2km) across, what the 21 transmitted pictures revealed was a crater-ridden surface pockmarked across an ancient, debris-strewn landscape. No irrigation channels, no crop fields, no towns or cities and very little reason to believe that there was any life at all on the planet. Over the next decade, fascination with Mars changed from one charged with the possibility of advanced life to one in which humans could find a new world to populate.

That possibility has received broad support among space advocates, futurists, and science fiction writers alike and it was that which stimulated Elon Musk to return to answer the bigger question ignored by H G Wells, but first asked by von Braun: could we be the first Martians? Musk became an early advocate of humanity's planetary migration in an age when environmental concerns had reached a peak, discussion about climate change, the consequences of uncontrollable nuclear war and the environmental destruction of life on Earth changing the way that a sustainable space programme could be imagined and justified.

And he is serious about that, a magnetic rationale for a new age of young scientists and engineers disillusioned with Earth-scavenging technology and the profligate pursuit of economic growth at the expense of the environment. These themes are common on college campuses across the United

ABOVE • The Gateway will be a way station in orbit around the Moon, here imagined by an artist receiving an Orion spacecraft at right. (NASA)

ABOVE • *In the late 1950s, Wernher von Braun proposed orbital stations, Moon bases, and Mars landings in sequence as the paradigm for future space exploration. (NASA)*

RIGHT • *A prophet of future space objectives, rocket scientist von Braun (right) worked with Walt Disney to promote space exploration as the next major goal for humans. (NASA)*

RIGHT • *In 1965 NASA obtained the first pictures of Mars showing a pock-marked surface very different to the one many believed supported some form of life. (NASA)*

States, and the message had been growing that responsible management of the Earth should be a template for future action, with the caveat that weapons of mass destruction are a credible threat to life on Earth. Time to press pause and reset the future for the human race, say advocates of a new future.

But how can that be achieved? Is it realistic to imagine a time in the foreseeable future when tens of thousands of emigrants would leave Earth for good and settle on another planet, guaranteeing the survival of the human race in the event of environmental destruction, global nuclear war, or the attentions of an impacting asteroid? Elon Musk thinks so, and that belief has provided the incentive for his very personal project funded by his own bank balance. Time to take a closer look at Starship.

Spaceships to Mars

Wernher von Braun imagined a fleet of giant spaceships departing Earth on flights of discovery to the surface of Mars, most of the structure, propellant tanks and rocket motors being expendable. Musk wants to go to Mars in a different way by developing the Starship rocket with two stages, each of which are totally reusable and with an initial payload capacity of 100 tonnes to low Earth orbit. Initially, SpaceX referred to the entire vehicle as Starship,

but latterly that has come to refer to the second stage only with the first stage known as Super Heavy. Starship will operate as an independent spaceship and a variant has been selected as the Human Landing System for NASA's Artemis Moon programme, as described in the previous chapter.

Initially, SpaceX looked to develop a completely new and more powerful rocket motor for both stages using liquid hydrogen and liquid oxygen propellants rather than the kerosene and oxygen for Merlin motors and bearing the name Raptor. By 2012, with considerably more people working on the new engine, SpaceX changed fuel and oxidiser to a mix of methane and oxygen with almost as high an energy output, propellants which Musk said might be available on Mars to process into rocket fuels for sustainability. Four years later, SpaceX received a contract from the US Air Force for development of Raptor as potential upper stage motors for the Falcon 9 Block Five and Falcon Heavy.

An early exponent of new technology, SpaceX used 3D printing for 40% by weight of Raptor components, including turbopumps and injector plates and soon would also incorporate a new Inconel superalloy which had been developed in-house. Several different Raptor engine designs were prepared with an optimised arrangement assigned to early Starships. While not providing as high an energy output as hydrogen, as a fuel methane has advantages in that it is denser and therefore requires smaller tanks for a given thrust and burn duration. Moreover, at -258.7ºF (-161.5ºC) the higher boiling point of methane compared to hydrogen at -423ºF (-253ºC) adds fewer structural demands on tank insulation to maintain cryogenic temperatures and that in turn reduces the weight of the tanks.

Three versions of Raptor are currently in production or development, Raptor 1 having a thrust of 408,000lb (1,814kN), Raptor 2 with a thrust of 507,000lb (2,255kN) and Raptor 3 with 593,000lb (2,637kN). In development testing, SpaceX believes that it can increase the output of Raptor 3 to a thrust of 727,650lb (3,236kN). Super Heavy would have 33 Raptor 2 motors at its base with a lift-off thrust of 16,700,000lb (74,281kN). This is considerably more than twice the thrust of the storied Saturn V used for Apollo missions, mostly to the Moon and for launching the Skylab space station into Earth orbit. For SpaceX, it is a very big step up, Super Heavy having more than three times the thrust of Falcon Heavy.

The Starship upper element has six Raptor motors with three used to decelerate the stage as it re-enters the atmosphere and conducts a braking manoeuvre for final descent to a soft touchdown on extendable legs. With considerable increase in thrust during vacuum firings, the Starship engines have a different configuration and skirt

LEFT • SpaceX is committed to developing the rockets and the spaceships to make Mars trips routine, starting from Boca Chica, Texas, where investment in a resort development comes complete with Tiki Bar. (Lars Ploughmann)

design to make the greatest possible use of that opportunity for the upper stage motors to produce the maximum thrust outside the Earth's atmosphere. This is the only significant difference between the design and manufacture of the separate sea-level and vacuum motors.

Almost all the manufacturing and assembly criteria focuses on the maximum use of common parts and assembly techniques, further reducing production costs. Raptor production takes place at Hawthorne, California, with tests conducted at McGregor, Texas, as it was for the Falcon 9 and Falcon Heavy programmes. A considerable amount of Raptor engine development involved government facilities at NASA's Stennis Space Center, initially set up in 1961 as the Mississippi Test Facility, alongside the Pearl River between Mississippi and Louisiana. It was there that rocket motors developed for the Apollo and Shuttle programmes were tested.

There were no existing facilities which could accommodate the Super Heavy/Starship launch system so the search began for a suitable location where rockets could be assembled and launched. The task was made difficult by the regulation that flights must not take place over populated areas for fear that debris from an explosion could fall to the ground and endanger human life. Several places were considered, from Florida to California and from Alaska down to Georgia but a final string of agreements was signed by state and federal authorities for a location at Boca Chica in Cameron County, Texas, close to the city of Brownsville. Adjacent to a nature reserve, special restrictions were placed on SpaceX for access to those locations. For launch paths, Super Heavy/Starship flights are directed southeast over the Gulf of Mexico and out across the Atlantic Ocean, any debris from an explosion falling in the water along surface corridors for which maritime vessels have been given prior warning.

As the launch cadence picks up tempo, SpaceX expects to significantly expand its Boca Chica facility, now named Starbase, but it is already having a significant impact on the local population. Recognising that sending the world's biggest rockets into space sits uncomfortably with a wildlife and wetlands area, Musk offered dispersed residents the chance of selling up. At first the majority refused, then changed their minds as the activity become more intense. Others still refuse to leave, many of them looking out for the wildlife, migrations of birds now not so frequently seen at Boca Chica, as the rocket men and women grow in number, louder in their activities and more disruptive to the natural ecology.

But people are also moving in, fans of SpaceX, enthusiastic supporters of interplanetary travel and others who recognise

that something extraordinary is happening there. Something that is seen nowhere else on Earth and that alone has brought a steady and consistent immigration of those simply wanting to be close to history, selling up and moving in on this otherwise quiet and vacant space where the Rio Grande meets the blue expanse of the Gulf of Mexico.

Perhaps the line where the present meets the future is hard to live with, as one resident noted: "I get a knock at the door at 10 o'clock at night, which is already shocking if you're not expecting someone. And there's the sheriff at my door handing me a notice that they're going to be testing between three and four am. And", she paraphrases, "it's recommended that you, and if you have any pets, shouldn't stay here". Surrounded on all sides by wildlife sanctuaries, even the birds are now leaving.

Big Stakes

The pace of flight test preparation at Boca Chica is unprecedented and at a scale that NASA would find impossible to match. Already, SpaceX has built 17 Super Heavy boosters, the first appearing in September 2020 and the most recent in March 2024, with many more to roll out at an increasing tempo. Since December 2018, 38 Starships have been assembled and used for various tests and preflight preparations and the fabrication pace is picking up there too with additional facilities being constructed on site supporting a 24/7 schedule.

Under a nominal mission, the Super Heavy booster ignites all 33 Raptor engines, which are mounted in three

BELOW • Development of the powerful Raptor engine (right) for Super Heavy/ Starship rockets at Hawthorne, California, alongside a Merlin engine from the Falcon programme. (Brandan De Young)

concentric rows, only the outer 20 not having gimbal capability for controlling the orientation of the stack. The booster stage shuts down all but three of the centre engines which throttle back to 50% thrust until they stop firing at two minutes 39sec, followed two seconds later by ignition of the six Raptor engines in Starship and stage separation.

The booster fly-back burn begins at two minutes 53sec and lasts for 54 seconds, re-igniting at six minutes 30sec for final deceleration prior to touchdown at six minutes 48sec. The launch gantry has scissor-arms which close in around the booster as it slowly descends back down to the launch pad and clasp it at a strong-point just below the grid fins at the top of the stage. When technicians ensure that all is safe at the pad below, the arms lower the stage back down for another launch. Meanwhile, Starship continues on into orbit and shuts down at eight minutes 33sec.

The first attempt at reaching orbit occurred at 8.33am local time on April 20, 2023, with several thousand spectators in public areas off site. At ignition, Booster 7 lost three of its 33 Raptor motors with two more going out at 29 seconds when a small fire broke out in a hydraulic power unit but when the stage shut down at two minutes 49sec, Starship 24 failed to separate and as the stack tumbled it broke apart and exploded. A flight-termination system failed to work and the assembly reached a maximum altitude of 24 miles (39km) before shattered fragments began raining down on the sea below. It had been intended to send Starship around the world for a simulated 'landing' off the coast of Hawaii.

The second attempt occurred on November 11, 2023, when Booster 9 and Starship 25 roared into the sky over Boca Chica. The flight had been held back after serious damage at the launch pad from the first launch required a re-think on its design, and the regulatory authorities had conducted analysis of the failed attempt. This time the Super Heavy booster behaved as required and Starship separated on time, failing within a few seconds of its

ABOVE • Variants of Raptor 2 with vacuum motor (right) and two sea-level motors inspected at NASA's Stennis Space Center. (NASA)

RIGHT • The Boca Chica rocket factory is growing in size, area and capability, preparation for a fast-paced launch programme which will be supported by a development involving employee homes and visitor resort attractions. (Steve Jurvetson)

intended velocity, although reaching a height of 92 miles (149km) before falling back through the atmosphere. This time the planned flight path for Starship was shorter but it disintegrated in a flaming trail through the atmosphere north of the Virgin Islands.

The third flight test on March 14, 2024, achieved most of the planned objectives but two of the three Raptors on Booster 10 shut down during the terminal descent into the sea and it broke up at a height of 1,650ft (503m). Starship 28 continued on around and started its re-entry from a maximum altitude of 145 miles (234km) before it began to roll uncontrollably and break up in the atmosphere. Again the FAA intervened after the flight to investigate the mishaps, while SpaceX sought to make minor changes for the next attempt which was planned for the second quarter of 2024. Never an organisation to waste time deliberating, SpaceX reiterated its former complaint that the FAA was taking too long to make its recommendations known and reissue a flight licence.

The challenges for SpaceX and its Super Heavy/Starship programme are great but the reward is unprecedented. Apart from using it to support manned landings on the Moon for NASA's Artemis programme, it is also the launcher of choice for sending up vast quantities of Starlink satellites at one time. Forecasts by independent analysts project a revenue of $6.6bn for 2024 which dominate a market where major geosynchronous satellite operators SES and Intelsat combined will obtain a revenue of only $4.1bn. The faster SpaceX can launch Starlink satellites the more it can grow the system, already reaching 2.7m subscribers in 75 countries with more than 3,000 of its 13,000 employees assigned to the programme. In comparison, NASA employs 18,000 people across all its sectors.

Starship is key to many existing programmes and will support the larger requirement of SpaceX in commercial applications, selling launch services to companies and organisations looking to place heavy payloads in orbit. It gains value from the existing inventory of Falcon rockets and of Falcon Heavy, which has also attracted orders from NASA. These are valued precursors to the bigger challenge of Starship and its built-in flight frequency of two or three launches a day from the same pad.

The very existence of a recoverable, soft-landing capability has pushed the Department of Defense to ask SpaceX to think about leasing Starship to the US Army for delivering very large

quantities of freight and cargo to any location on Earth. In fact, one Starship could send 150tonnes of freight to anywhere around the world within one hour from launch to landing, a load equivalent to two C-17 transport aircraft which would take almost a day to reach a distant destination.

It used to be impossible to think of launching a second rocket from the same pad in under several weeks; SpaceX is already achieving an average of one flight every three days from two locations (Vandenberg Air Force Base and Cape Canaveral) and looking to increase that significantly over the next several years with Falcon 9 and Falcon Heavy. A largely unrecognised transformation has taken place with frequency of launch, which was previously constrained by demand. As that has increased, and launch prices have tumbled under SpaceX, the frequency of launch is no longer

ABOVE • Super Heavy on the launch pad with grid fins extended and an approaching Starship to which it will be mated. (SpaceX)

LEFT • Starship is brought to the top of the Super Heavy booster for attachment, technicians giving scale to the 30ft (9.1m) diameter of the two stages. (SpaceX)

constrained by physical factors such as pad refurbishment after each flight and the ability to process several rockets of the same type concurrently. That capability has become a benchmark for SpaceX operations.

So it is that the ambition to fly frequently, at least several times a day, could be key to getting future prices down to levels equivalent to a ticket on a passenger-plane today. If that does become possible, the 680 people that have ever flown in space could be lifted into orbit aboard two Starships on the same day. Only a few decades ago, this would have been the stuff of science fiction, but it is perfectly reasonable to make this happen within the next decade. Perhaps even before the first missions set off for Mars.

Red Dragon

The way commercial contenders could take over big goals, even national objectives, is fundamental to how humans could get to Mars, even to colonise it as Elon Musk wants to see happen in his lifetime. His dream of that goes back more than a decade to the early days of his entry into rocket design and it really began in a study he conducted for NASA in 2011 on ways to examine surface rocks for biosignatures indicative of primitive life.

Never short on ideas, when the Dragon capsule for cargo to low Earth orbit was developed using contracts from NASA, Red Dragon was proposed as an uncrewed carrier for science experiments to do surface analysis on samples retrieved from a depth of 3.3ft (1m) by drill. Musk wanted it as a pathfinder for crewed missions to be followed by a more advanced spacecraft. It was proposed as a NASA science

mission in 2013 and again in 2015 but it was not selected for development. Had it happened, it would have been the first public-private partnership mission to another planet.

In 2014, a hybrid concept proposed a version of Red Dragon that could be used to return Mars samples to Earth for analysis. The flotilla of Mars-bound spacecraft over the preceding decade had brought urgency to the need for scientists to get their hands on rocks from the Red Planet and conduct advanced examination, again looking for signs of life. No amount of automation could replace the large, heavy, and expensive laboratory equipment on Earth necessary for the next level of study. SpaceX proposed Red Dragon could obtain samples, launch them off the surface and deliver them back to Earth in a recoverable capsule. That too was never implemented.

When the universal Dragon 2 spacecraft came along, in 2016 SpaceX proposed a flight launched by Falcon Heavy to achieve the same result and NASA showed some interest, suggesting that two Dragon 2s should be launched during the 2020 window, the second for redundancy. There was now significant interest, the New Space way of joining in partnership with private industry to accomplish expensive missions beginning to seed initiative across the spectrum of space endeavours and not just in manned flights to low Earth orbit. But it never happened as NASA turned its attention to a different way of getting samples back to Earth. Paradoxically, it was not the end of such ideas, however.

In a classic case of concept migration, the soft-landing technology for Falcon 9 and Falcon Heavy was instrumental in proposed Mars lander design. The 2016 sample-return proposal had sponsored development of a Mars landing module for Red Dragon 2, together with a heat shield uniquely adapted to the requirements of entering the atmosphere of Mars. Wanting to avoid the use of parachutes for slowing down on entry, SpaceX calculated the aerodynamic drag of the capsule alone could reduce speed from 13,000mph (20,917kph) down to the point where the Super Draco abort motors could take over and reduce vertical velocity to a soft touchdown. With as little as 4,190lb (1,900kg) of propellant, the rocket motors would have sufficient energy to achieve that.

The mathematics and the engineering were sound and had the value of allowing access to a wider range of mission objectives using techniques crafted for that initial objective. But at the time this proposal was being fielded, the Super Heavy/Starship combination was reaching the final concept stage. As previously noted, the BFR, as it was then called, was proposed for highly advanced missions and named the Interplanetary Transportation System (ITS). At this time it was envisaged to have 42 Raptor engines in the first stage (Super Heavy) delivering a lift-off thrust of 29,000,000lb (128,992kN) with nine Raptors in the upper stage (Starship).

The Mars Colonial Fleet was envisaged as providing transport for initially 100 and later 200 people on each launch, the first stage in a routine schedule for moving large numbers of people to the Red Planet. From the outset, Starship was designed along the lines of an ocean-going liner with propulsion in the rear and passenger space with spectacular views forward, capped by a panoramic window section for stunning views of the universe on a scale no other spacecraft or spaceship design could offer.

Key to achieving this was the ability to fuel a Starship in orbit before departing for Mars, as there was insufficient propellant after it had placed itself in space to propel it out of Earth orbit and on to the Red Planet. This had never been attempted before yet it was crucial to the concept and considerable design and engineering effort was applied to being able to accomplish rendezvous and docking of a Starship tanker with a Starship passenger-carrier.

The difficulties in achieving that are self-evident. Moving cryogenic propellants around on the ground is difficult enough but trying to achieve that in orbit compounds the challenge. Nevertheless, it is fundamental to what any programme such as this would have to accomplish and is the inevitable and unavoidable next step in transitioning from an Earth-based economy to a multi-planetary society.

In the very early days of the space programme, Wernher von Braun's Saturn rocket team had proposed such a possibility and SpaceX re-examined that concept 60 years later with a plan to dock them side-by-side for connecting propellant transfer connections from the tanker to the passenger-carrier. It is this technique that will be used with the Starship Human Landing System supporting NASA's Artemis lunar landing programme.

Precursor Steps

The Super Heavy/Starship programme has eclipsed all prior applications of Dragon capsules for landing on Mars, using scale to cut launch costs and lower the projected funds required to begin establishing bases and then permanent habitations on Mars. From such facilities would come several separate communities on the Red Planet, growing in size and connected by intercity transport links using hopper-craft 'flying' between locations while viewing surface features below.

But such dreams are made real only through the perfection of new capabilities in technology and engineering. As it was with the development of human civilisation on Earth, from the star logs of Babylon, the engineering accomplishments of the ancient Chinese empires, to the science and mathematics of the Greek island city-states, human progress is built on creativity and ingenuity. In the meantime there is much to do and the

proven reliability of Falcon and Falcon Heavy has attracted more attention from NASA.

To support the objectives of the space agency's Artemis programme in establishing a permanent base for scientific research on the Moon, an efficient and effective support infrastructure will be essential, much as it has been for maintaining the International Space Station in low Earth orbit. NASA plans to de-orbit the station in 2031, a colossal task in itself and one which will be conducted with precision. Not all of it will burn up in the atmosphere and much of the structure will survive re-entry to impact the ocean. It will be brought down in a pre-planned zone anywhere between 51.6° north and 51.6° south of the equator. That is a job for which the only affordable solution may be to use a commercial launch vehicle carrying a large propulsion module.

Long before that takes place, NASA will have begun assembling another space station, much smaller than the ISS and orbiting the Moon in a path that takes it from a distance of 1,900 miles (3,000km) above the surface out as far as 43,000 miles (70,000km), making one full revolution of the Moon every seven days. Known as the Gateway, it is a truly international effort, with modules and equipment from the United States, Europe, Japan, Canada, and the United Arab Emirates. This is a major step forward in studying space science and the physiology of astronauts outside the protective envelope of the magnetosphere, which insulates all life on Earth from harmful radiation.

The Gateway is intended to be a focus for wide ranging scientific investigations but also to serve as a terminus and relay point for permanent bases on the Moon. It will have facilities supporting up to four people but it will not be permanently

ABOVE • An uncrewed Red Dragon imagined completing a soft-landing on Mars for initial sampling, as proposed by SpaceX in 2016. (SpaceX)

BELOW • Super Draco engines bring Red Dragon to a soft touchdown in the early stages of a Mars settlement, a concept proposed by SpaceX but never adopted. (Kevin Gill)

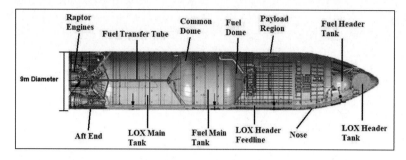

Raptor Engines • Fuel Transfer Tube • Common Dome • Fuel Dome • Payload Region • Fuel Header Tank • 9m Diameter • Aft End • LOX Main Tank • Fuel Main Tank • LOX Header Feedline • Nose • LOX Header Tank

ABOVE • The internal structure of Starship with spacious volume for cargo or passengers, up to 100 people being comfortably accommodated in its forward section. (FAA)

BELOW • Starship will support development of a lunar base and provide heavy-lift cargo replenishment from Earth for scientists and personnel at the Moon. (SpaceX)

occupied, operating as a docking station for NASA's Orion spacecraft bringing people from Earth to transition to the SpaceX Starship lunar lander in which they will go down to the surface. Over time it will serve as a holding station for rocks from the surface, samples probably returned to Earth in Falcon Heavy rockets, for which there is already a contracted role.

NASA has tasked SpaceX with delivering the first two elements of the Gateway, the power and propulsion element and the habitation and logistics outpost, presently planned for launch in November 2025 on a Falcon Heavy. To this will be attached further modules over time, although it is not expected to support the first Artemis Moon landing. The Gateway orbit will be at 90° to the Moon's equator, with the low point of the orbit passing across the North Pole before it swings much further out over the South Pole. Astronauts will be able to stay in the Gateway for up to 30 days and this polar-orbiting path has advantages in that the time spent passing across the far side of the Moon is minimised.

Any development of either the Gateway or surface bases will require logistical support and the template of unmanned cargo flights now routinely flown for the International Space Station can be slotted in to flight planning supply trips to this more distant location. It effectively moves out the logistical operation from Earth orbit to lunar orbit, creating a transportation route between Earth and Moon in what scientists call cislunar space. It is

therefore a logical move for NASA to have held discussions about that role, for which Falcon Heavy will be particularly well suited due to its large payload capacity. And it will be supporting both ISS and Gateway logistical requirements in the overlap while both are still functioning.

There is perhaps an additional role which SpaceX can play. For several years, NASA has been mired in debate over funding for a Mars sample-return mission and we saw earlier how that had initially involved SpaceX and its proposed Red Dragon, to no avail. On February 18, 2021, the Perseverance rover landed on Mars and began an intense geological survey of its vicinity, to date having traversed more than 16 miles (25.7km), sampling, probing, and collecting material which it has cached into tubes, up to 43 being eventually filled, sealed and left for collection at some future date.

The cost of mounting a mission to retrieve those is too costly for Congress to approve funding and NASA is looking for bright ideas that can slash the money required to do it. It just may fall to the Super Heavy/Starship launcher to perform such an operation but to do so would require considerable change to its capabilities. Low launch costs are only one small factor in reducing the price of such an advanced mission. But if that were to happen it would provide a fitting precursor to the role Starship would be called on to perform in the future where the enormous carrying capacity of this rocket is framed as the bedrock of a migration from Earth to Mars.

One thing is certain. With so many plans and such a diverse range of programmes and projects already funded and in work, many more people will be required to support an industry which already employs 500,000 people and is worth around $500bn annually and expected to reach $1trillion by 2030.

So, just how rewarding is it to choose a career in space engineering? In financial terms a space engineer in the UK can get an average annual salary of £35,000, or twice that in industry positions attracting managerial roles. The average salary for a SpaceX engineer is $138,000 (£110,000), or

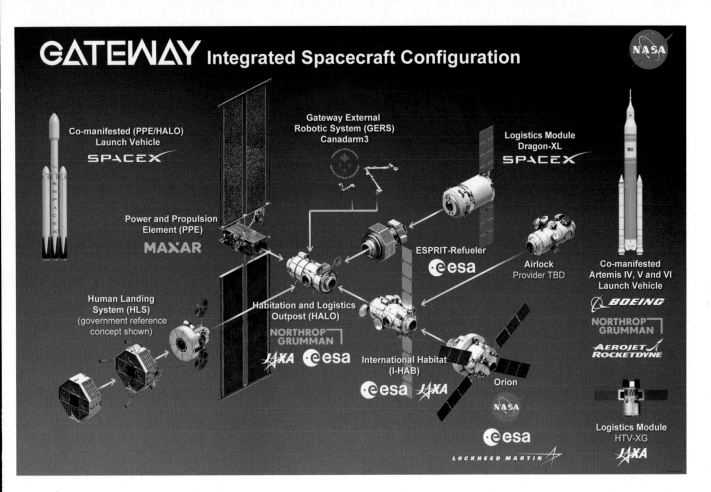

GATEWAY Integrated Spacecraft Configuration

Co-manifested (PPE/HALO) Launch Vehicle — SPACEX

Gateway External Robotic System (GERS) Canadarm3

Logistics Module Dragon-XL — SPACEX

Power and Propulsion Element (PPE) — MAXAR

ESPRIT-Refueler — esa

Airlock Provider TBD

Co-manifested Artemis IV, V and VI Launch Vehicle — BOEING, NORTHROP GRUMMAN, AEROJET ROCKETDYNE

Human Landing System (HLS) (government reference concept shown)

Habitation and Logistics Outpost (HALO) — NORTHROP GRUMMAN, JAXA, esa

International Habitat (I-HAB) — esa, JAXA

Orion — NASA, esa, LOCKHEED MARTIN

Logistics Module HTV-XG — JAXA

NASA

---

around \$108,000 (£86,000) at NASA, with the inevitable wide variations according to age, experience and job position.

The time and cost it takes to achieve the qualifications to do those jobs is no greater than other equivalent positions in different industries, but there is a wide range of ancillary positions in support roles and associated industries, attracting a wide range of differently talented people. SpaceX is generally considered to be a reputable employer, with average salaries around \$97,000 a year across all the work categories from clerical to mission management roles, and with a high rating from anonymous employee reviews.

Professional astronauts launched to the ISS receive salaries based on government pay grades when employed by national space agencies, or by negotiation when employed by private companies such as SpaceX. NASA pays its astronauts between \$81,000 and \$105,000 per year according to the relevant grade for which they are qualified, while the European Space Agency pays relevant salaries based on national standards. The French pay €70,000, the Germans €66,000 and the British around £54,000. However, EU law does not require the citizens of member countries to pay income tax when employed by ESA so British astronauts working for private companies will pay that.

Much of the US space programme managed by NASA is operated from, and controlled by, private companies with vertical integration, organisations that can make quick decisions, process new designs at pace, and produce results much quicker than the heavy bureaucracies that constrain government departments. The transfer of design and production of key launch vehicles and spacecraft to the commercial sector was a hard decision to make and a lot of it came down to young entrepreneurs providing workable and highly cost-effective solutions to problems faced by diehard managers in NASA, at ease with handing over responsibility for innovative solutions. It worked.

However, with commercial companies carrying astronauts to and from the International Space Station for NASA, has it all been worth it? Economically, no; politically, yes. Between 2006 and the present, NASA paid the Russian government around \$4bn for 71 seats on Soyuz spacecraft, a figure far beyond what it had expected to pay before the extended delays to getting private US operators up and running. Balanced against that is the more than \$6bn spent to date on paying for seats on US operators SpaceX and Boeing.

While retaining overall direction of policy and national goals in space, NASA would like to offload responsibility for operating private space stations based on the same model as cargo and crew development programmes for the ISS. Undoubtedly, the immediate future is locked into a learning curve, developing the machines, the tools and the people to establish a permanent research station on the Moon, and in so doing to develop the means by which the first expeditions to Mars can take place. In doing so, it may begin a new phase in human and societal evolution. One in which humans become a three-planet species, living, working and journeying between Earth, the Moon, and Mars.

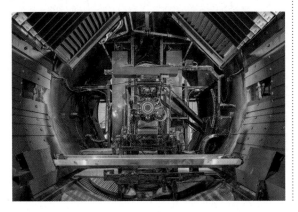

ABOVE • In support of NASA's international Artemis Moon programme, SpaceX will use Falcon Heavy to send the first modules to build the Gateway where, on later missions, Starship landers will receive astronauts for descent to the surface. (NASA)

LEFT • A section of the power system for the Gateway's Advanced Electric Propulsion System gets a test at NASA's Glenn Research Center. (NASA)

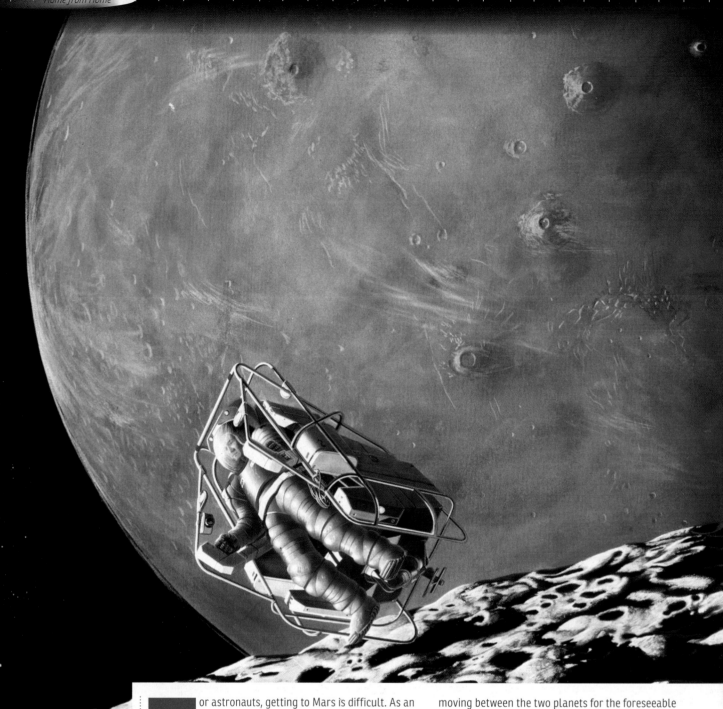

ABOVE • NASA imagined a range of different missions to Mars, some of which would have stopped at either one of its two moons. (NASA)

RIGHT • Geologists support engineering contractors by studying rock formations, producing maps to identify natural resource materials. (NASA)

For astronauts, getting to Mars is difficult. As an outer planet to Earth there are times when the two worlds are moving past each other on the same side of the Sun and only 48 million miles (78 million km) apart. Because Mars moves more slowly in its solar orbit than the Earth, there are times when the two planets are on opposite sides of the Sun, a distance of 250 million miles (401m km), before the Earth speeds round again and catches up and overtakes it once more on the inner track.

There are consequences for that which seriously impact missions where astronauts rely on communications and, because of the difference in distance, it can take between 4.3 minutes and 22 minutes to send a signal between the two planets. In the most extreme case, it would take 44 minutes to ask a question and receive an immediate answer. This means that activities and decision-making would have to be made at the surface of Mars and the ability of the spacecraft and the astronauts to control their affairs and solve problems, without consultation with Earth, would bring a new and perhaps worrying level of independence to the crew.

Flights to Mars can only be undertaken once every 26 months due to the alignment of the two orbits. Using conventional propulsion, which will be the only way of

moving between the two planets for the foreseeable future, it takes about nine months to get there. Once on the surface astronauts would have to wait three months to start back, also taking about nine months – 21 months overall. Reducing the transit time is achievable with nuclear propulsion and solar-electric propulsion, but the size of such systems has not been tested at scale and with sufficient thrust to offer an early solution.

THE FIRST MARTIANS

Living and working on the Red Planet

Surviving on the surface of Mars would be no mean feat, the thin atmosphere which has less than 1% the pressure at the surface of the Earth and being almost completely carbon dioxide it is toxic to humans. Moreover, due to its distance Mars receives only 43% the amount of sunlight that reaches Earth and temperatures range from -166ºF (-110ºC) to a high of 95ºF (35ºC) at the tropical equator, with the mean planetary average being -80ºF (-60ºC). There is no standing water, although it is clear that water and carbon dioxide ice exist below the surface and at the poles. Moving around is more comfortable than on the Moon, with has one-third Earth gravity compared with one-sixth on the lunar surface.

The atmosphere itself can pose hostile challenges, winds of up to 100mph (160kph) are common and dust storms regularly envelop the planet on seasonal variations. It was these planet-wide changes on Mars that gave some astronomers the notion that surface features were made by vegetation and crops planted by intelligent beings. The surface of Mars itself varies from place to place, giant canyons extending a distance equal to the continental breadth of the United States, and volcanoes up to three times the height of Mount Everest, while other places are like river deltas strewn with high sand dunes.

Moving around on the surface brings its own difficulties due to the different types of terrain and the abundance of rocks. Atmospheric erosion has far less effect than on Earth due to the much lower pressure which has a less abrasive effect. Mars rovers have experienced the consequences of that where wheels are cut and ripped by sharp rocks. Vast dust bowls are not uncommon and huge caves challenge giant rock cavities for openings that threaten vehicles, some of these extending to great depth. Navigation across the surface is difficult due to the lower light levels and virtually impossible in the dust-storm season.

Several companies have looked into the challenges of sustainability, primarily the need for food on a world which has no known living organisms. It will be necessary to grow food from seed and that initiative on the International Space Station has already produced positive results. However, the first visitors from Earth will need large quantities of supplies to get

ABOVE • Commercial contractors on Mars assemble work stations and living quarters to start surveying for resource materials. (NASA)

BELOW LEFT • Initial landings on Mars will be supported with primitive dwellings to begin the process of building a habitable station. (NASA)

BELOW RIGHT • Over time, greenhouses would be constructed for large-scale food production supporting a self-sustaining community less dependent on Starship cargo tankers from Earth. (NASA)

greenhouses up and running and that would be possible with Starship and its large cargo capacity.

Based on consumption at the space station, each astronaut would require 2,542lb (1,153kg) of food for a 21-month round-trip, perhaps five tonnes for a crew of four or 11 tonnes for a crew of 10, which some consider to be optimum for such a special expedition. But this does not account for the amount of water required, which may be recycled and therefore on a more sustainable basis, nor does it address the amount of exercise equipment needed to offset the effects of 18 months weightlessness and three months in one-third gravity.

Health is an issue which has been examined and based on experience with the International Space Station, there are basic services which individual crewmembers could be trained to provide, even some invasive surgery such as dentistry or other non-life threatening complaints that could erupt over that period. It is likely that a physician would be assigned to the crew and provide skilled medical services, this being a requirement in several studies during the early space programme. The need to monitor health is made all the more important by the exposure to radiation which Earthlings and even occupants of the space station are protected from by the Earth's magnetosphere and its doughnut-shaped magnetic field.

The psychological effects of deep-space voyages are believed to be largely unknown. Nobody has been away from Earth for that long and completely out of visual range of the home planet. The so-called overview-effect, where the presence of the Earth below in all its panoramic

magnificence brings comfort to occupants of the space station, is considered an important relief. Without that, nobody knows how individuals might react, where Earth is merely one among the stars and the Moon can no longer be seen. Nor how they might respond to a major problem threatening their lives where technical failures could bring a very real possibility of dying in space.

On the surface of Mars, life would have to exist within a bubble of air and a shirt-sleeve environment without which it would be impossible to survive. The most difficult expeditions would be those in the initial stages of settling the planet with people, before the infrastructure had been set up to manufacture life-giving air, recycle water and grow food. People would have to exist in structures built partly below the surface, if only for protection from radiation, and while artists continue to show transparent canopies covering surface colonies, in reality these would be difficult to insulate and to make impervious to micrometeorite impacts and the occasional effects of a wind sand-blasting the surface.

Paying the Piper

How is all this going to be achieved, who will do it and where will the money come from? Governments began the space programme paid for by taxpayers supporting national efforts for power and prestige built on the engineering developed for guided missiles and ballistic rockets. That period began during the Cold War in the late 1950s and extended into the early 1980s as a variant of proxy conflict where the battle for prestige was played out in space.

Then the second phase began in which international cooperation replaced the jingoistic search for technological virility as a symbol of national pride, with links forged across the ideological divide to build giant telescopes placed in orbit, expeditions to planets and asteroids, and cooperative programmes for space-based medical research and the general scientific exploration of the solar system. In a way, the first two phases retain the connection between technical achievement and national pride but that was about to change.

Starting in the new century, entrepreneurs and billionaires sought fulfilment not through conspicuous consumption of material possessions but rather through a different ideology to that which previous governments alloyed to their taxpayers. One in which a complete change of direction opened new markets for commercial gain that provided the funds to pay for projects so big that they were beyond the means of government, even international partnerships, to pay for. By developing a cheaper way of launching satellites, the number of companies finding it profitable to invest in space applications such as navigation, TV, voice-communications, and internet services grew. That brought more money to the New Space entrepreneurs and venture capitalists to grow the sector, expand opportunities and bring returns to do things governments are no longer capable of.

Mars Greenhouse

NASA bridges all three stages, retaining a sense of responsibility for its Congressionally-mandated role of keeping the United States 'pre-eminent in space', extending shared development of new programmes with partner countries around the world and leasing services from commercial space companies to do the things it can no longer afford to pay for. In its aspiration to further research with robot landers and rovers on the surface of Mars, NASA is unable to pay for the vehicle to retrieve samples it had previously sent a spacecraft to collect. Over the last several decades, with declining budgets the agency is no longer able to get the funds to fulfil its objectives.

Following Mercury, Gemini, Apollo, the Lunar Module, and the Shuttle, NASA is developing its sixth crewed spacecraft. But for the first time it is a partner deal with the European Space Agency, where only half is built in the United States. The Gateway operating as a way-station around the Moon is an assembly of modules from partner countries around the world. But the lifeline to the International Space Station, the launch and servicing of the future Gateway and the spacecraft that will carry astronauts down to the Moon are all from private, commercial companies who own the hardware. And now, those same commercial companies are flying the rockets and the spacecraft that could carry humans to Mars for the first time.

For as long as it has existed, NASA has looked to Mars as the ultimate goal for its human space flight programme. The Russians aspired to that goal too, before the Soviet Union collapsed in 1991, Russia ran out of money and its government lowered aspirations and found other things to do. China pledges that it will have people on the lunar surface by 2030 and is likely to turn next to Mars, although the resources on the Moon may tie it up there for several decades to come. Which is why so many companies in the West are working to develop equipment that could start to mine there for rare-earth minerals vital for the age of electronics and advanced AI systems needing exotic materials.

It is in this sector that the funds may be generated for commercial management of Mars missions, regulated by standards and conditions from government agencies and using the extensive facilities of organisations such as NASA for testing, tracking, and communications. Facilities which will also be available for space medicine and regulatory health standards for crew members. It would appear that gone are the days when even the richest governments on Earth can pay for the ultimate visit by humans to another planet in the Solar System. They have been trying to do that for more than 60 years, successive administrations promising unachievable goals with unavailable funds.

As to who will fund it, self-evidently the companies and organisations that aim to profit the most from investment in grand expeditions to new places in search of greater wealth and, in return, their own prestige. It was, after all, the way the world emerged from the Age of Enlightenment and reached out around the globe, girdling the Earth with seaways and trade routes to enrich the entrepreneurs and the billionaires of a former age, building an industrial revolution to power new possibilities and in so doing raise standards for the living and promises for the yet unborn. It is, after all, how Elon Musk thinks.

For those who decry the ever-surging race for human progress, pointing to the Earth and its environmental ills, unquestionably in dire need of restitution, there is no place for any of this. But those who follow Musk believe that humans need a place where they can begin over without the mistakes of the past, a new world in which to develop a better place to live and perhaps reshape through terraforming a hostile world into a second Earth flooded with life luxuriating in a breathable atmosphere and with seas and rivers.

The concept of terraforming has been around for a very long time and there are many who believe that such things will never be possible. But there are great minds today who support the idea of doing that and they have an alternative view of how humans will settle on Mars. Such possibilities for turning Mars into an Earth-like world are far in the future, perhaps several centuries. Before that can happen, the likelihood that SpaceX and Elon Musk's dream to put humans on Mars is possible within the next decade or so. There is no alternative or viable way of achieving that and if mining the Moon and supplying Earth industries with the materials they require will be as lucrative as many believe, there are many alive today who could make that first trip to the Red Planet. They would, after all, be the first Martians.

SHEPARDS
TO THE GLENN

A low-profile player seizes the initiative

Born in Albuquerque, New Mexico, in 1964, Jeff Bezos is regarded as the second richest man in the world with a net worth of around $205bn, only marginally ahead of Elon Musk. With degrees in computer science and electrical engineering, Bezos developed a knack for business during his time on Wall Street from 1986 until he founded Amazon in 1994, now the world's biggest e-commerce and cloud computing company. A lifelong enthusiast for space travel and the expansion of the human presence across the Solar System, he founded Blue Origin in September 2000, arguably the first of the billionaires who would make a commitment to a New Space goal of commercialised space exploration.

In some respects, Bezos would follow Richard Branson with a space-tourist attraction through his New Shepard rocket, named after the NASA astronaut Allan Shepard, who on May 5, 1961, became the first American to go into space in a Mercury capsule launched by a converted Redstone rocket. Bezos had great attachment to these early flights, inspired by events that had taken place several years before he was born.

The basic concept for New Shepard was a rocket capable of firing a reusable capsule carrying six people or cargo to an altitude of at least 100km, which is the internationally recognised demarcation line between the atmosphere and outer space. The capsule would be recovered by parachute and a solid propellant

BELOW • New Shepard lifts its first passengers on a rocket-ride to space, July 20, 2021. (Blue Origin)

rocket motor fired just before touchdown for final deceleration before a soft landing. The booster rocket would make a controlled descent and land vertically supported by four extendible legs.

This market is one also addressed by Virgin Galactic and its winged vehicles launched from beneath a White Knight Two carrier-plane, but the Blue Virgin process is a lot simpler, requires much less time from climbing aboard to experiencing the flight, and provides the passengers with a longer period of weightlessness. Moreover, it provides an opportunity for passengers to be internationally recognised as astronauts while the Virgin Galactic flights merely exceed the 50km line recognised by NASA and the US Air Force, which means that they are not true space flights.

Blue Origin began development with the Goddard test rocket powered by nine BE-1 rockets, each of which provided a thrust of 2,000lb (8,896N) from a peroxide propellant. With these, the rocket, named after Robert Goddard, the first man to demonstrate flight with a liquid propellant rocket motor in 1926, was launched on November 13, 2006. The rocket reached a height of 279ft (85m), returning to a perfect landing on legs about 25 seconds after lift-off. A very small step toward the development of the New Shepard people-lifter, it was a success for Blue Origin and gave Bezos his first public declaration of intent on social media. The third and last flight with this rocket occurred on April 19, 2007.

Unlike Richard Branson and never a flamboyant publicist, Bezos was quiet about his work which proceeded methodically and without fanfare, no marketing taking place and progress consistent if low profile. The design of the crew capsule went on apace and a successful pad abort test was conducted on October 18, 2012. Flight tests and launch activity were conducted from a 165,000acre (670km²) piece of Texas about 30 miles (48km) north of Van Horn and known as Corn Ranch, procured by Bezos in 2004. It is perfectly suited to the isolated work going on there to develop a commercial space tourism business. Its solitude fits well with the equally introspective Robert Goddard.

With a thrust of 110,000lb (489.2kN) and a burn time of two minutes 21sec from hydrogen and liquid oxygen, the BE-3PM rocket motor which would power New Shepard completed its test firings in April 2015. The first flight occurred on April 29, but a hydraulic failure caused it to collapse on returning to the ground after reaching a height of 58.1 miles (93.5km). It had carried the capsule RSS *Jules Verne*. Three New Shepard rockets were built during that year and the second completed five successful flights before it was retired in October 2016.

New Shepard 3 was a further improvement and would carry the capsule RSS *H. G. Wells* on development flights carrying only packages of experiments and no passengers. The first flight on December 12, 2017, was the first certificated by the FAA under new legislation which would regulate commercial, non-government rocket flights. It did fly commercial payloads for several minutes of weightlessness of value to researchers who pay for its use, and it also carried a test dummy, 'Mannequin Skywalker'.

New Shepard 3 was used for several development tests, but much was being learned in this phase of test-and-try, procedures and operating modes which did not always go as expected, although it did achieve an altitude of 66.4 miles (106.9km) on January 23, 2019. After successfully testing a new landing technology, it was destroyed on September 12, 2022, due to an incorrect in-flight abort command. Its successor, New Shepard 4 was ready long before this,

ABOVE • The Mk 1 Blue Moon spacecraft which will carry cargo down to the surface of the Moon. (Blue Origin)

LEFT • Jeff Bezos has developed Blue Origin into a company offering space tourist flights, a heavy-lift satellite launcher and is now building a Moon lander for NASA. (Blue Origin)

RIGHT • Launched on March 31, 2022, NS20 passengers (from left) Marc Hagle, Gary Lai, George Nield, Jim Kitchen, Marty Allen, and Sharon Hagle. (Blue Origin)

however, and was slated to be the first to carry humans into space for Blue Origin. It made its first flight on January 14, 2021, uncrewed and carrying test equipment to a height of 66.45 miles (106.9km). The second flight on April 14 was a pre-crewed test of procedures for a passenger flight, involving employees Gary Lai, Susan Knapp, Clary Mowry, and Audrey Powers, rehearsing all the procedures inside the capsule before leaving the capsule prior to launch.

The first flight carrying passengers to the edge of space was launched on July 20, 2021, the 52nd anniversary of the first manned landing on the Moon. Blue Origin ran an auction for one of the four seats. With a bid of $28million, the Chinese-born cryptocurrency executive and founder of TRON, Justin Sun should have been on board but had conflicting commitments and the seat went to runner-up Joes Daemon, the Dutch CEO of Somerset Capital Partners, who gave the slot to his son Oliver, at 18 the youngest person to go into space. Others on board included Jeff Bezos and his brother Mark together with Wally Funk, who had tried to become an astronaut when NASA was only recruiting men. She finally made it into space, achieving a height of 66 miles (107km).

Following a flight on August 25, 2021, carrying experiments, the second crewed flight was made on October 13, the first with six people, that number also flown on December 11, which included Alan Shepard's daughter Laura. Three six-person flights were made in 2022 before a flight carrying 36 payloads was launched on September 12 aboard the New Shepard 3 booster, which suffered a failure when its BE-4 rocket motor sputtered out and the abort system took over, carrying the capsule to safety. Following a thorough investigation from the FAA, flights resumed on December 19, 2023 when New Shepard 4 flew 33 payloads to a height of 66.5 miles (107km).

Crewed flights resumed on May 19, 2024, carrying Ed Dwight among the six passengers, the first black astronaut

RIGHT • Mário Ferreira (left) and Corby Cotton enjoying weightlessness while on the sixth crewed flight launched on August 4, 2022. (Blue Origin)

BELOW • Hardware coming together for the New Glenn launch vehicle scheduled to launch first in September 2024 carrying two small satellites to Mars. (Blue Origin)

who had been approved by President Kennedy in 1961 but never selected for flight. Now at the age of 90 he is the oldest man to have flown in space, next to actor William Shatner who had flown aboard the second Blue Origin passenger flight at the same age but younger by several months than Dwight when he flew.

On to a New Glenn

To date, Blue Origin has made $100m from the tourist and payload flights to the edge of space, flights which last little more than 10 minutes from launch to landing and which are attracting increasing levels of interest. But having launched 32 people to space, Jeff Bezos has already turned to much more ambitious plans, involving not only a new commercial rocket to lower satellite payload costs and compete with SpaceX Falcon 9 and Falcon Heavy, but also to put astronauts down on the Moon by the end of this decade.

Bezos has been funding development of a heavy launch vehicle capable of placing 99,000lb (44,900kg) in low Earth orbit or 30,000lb (13,600kg) into a geosynchronous transfer orbit to service the needs of satellite operators using that location for telecommunications. While New Shepard uses liquid hydrogen and liquid oxygen, the reusable first stage of New Glenn is powered by seven BE-4 rocket motors fuelled with methane and liquid oxygen, the same as the SpaceX Starship. New Glenn will deliver a lift-off thrust of 3,850,000lb (17,124kN), about half the thrust of Saturn V from the Apollo era. The expendable second stage has two BE-3 engines producing a thrust of 320,000lb (1,423kN)

Development began in 2012 and Blue Origin has signed a contract to use Launch Complex 36 (LC-36) at Cape Canaveral, originally two pads previously used for launching 145 Atlas space missions between 1962 and 2005, another example of where commercial operators are taking over pads previously used exclusively for NASA and US Air Force launches. Blue Origin expects the first flight of New Glenn to occur in September 2024 and that is critical because it will carry NASA's EscaPADE mission, two small spacecraft each weighing 1,200lb (540kg) to operate from a highly elliptical path around Mars. Blue Origin also has a contract with Amazon for 12 flights aboard New Glenn and an option of a further 15.

Its aspirations growing, Blue Origin competed in 2019 for a NASA contract to develop a lunar lander to operate with the Orion spacecraft to put astronauts on the Moon. It followed recognition that NASA alone could not afford to fulfil the obligation set by Congress to put astronauts back on the surface and develop a sustainable presence. Three organisations competed for that work, defined as the Human Landing System (HLS), which required the winner to conduct one uncrewed and one crewed demonstration Moon landing. Others would follow under a leasing arrangement. SpaceX offered its Starship at a bid price of $2.94bn, followed by Blue Origin at $5.99bn and Dynetics with a quote of $9.08bn.

On April 16, 2021, SpaceX was announced the winner, but Blue Origin contested that decision and lost. NASA had wanted to place contracts with two bidders, as it had with the commercial cargo and crew programmes for the International Space Station, but it had insufficient funds to do so. Later, under pressure from Congress and with a little more money, on May 19, 2023, NASA announced that Blue Origin would build a second lander to a very different design with an anticipated flight date around 2029 for a fixed-price contract worth $3.4bn. Known as Blue Moon, the lander will provide a back-up capability against failure to deliver SpaceX's Starship HLS.

The Blue Moon programme involves an architecture which includes a Mk 1 robotic lander capable of placing 6,600lb (2,993kg) on the lunar surface and a Mk 2 designed to carry four astronauts for missions on the Moon lasting up to 30 days. An uncrewed variant of the Mk 2 could deliver 44,000lb (19,958kg) for sustainable operations on the lunar surface. Both variants are designed for launch on New Glenn rockets. Design, development and fabrication of rockets and spacecraft takes place at Exploration Park alongside NASA's Kennedy Space Center, which helps integrate hardware, crew, flight preparation and launch within close proximity.

The plan for lunar landings with astronauts begins with orbital tanker fuelling operations so that the Blue Moon lander can be sent to the Gateway where it awaits the arrival of a crew carried by the Orion spacecraft. From there it takes the four-person crew down to the surface, leaving Orion at the Gateway. After a period of surface exploration, the Blue Moon spacecraft delivers the crew back to the Gateway for a return flight to Earth. There is considerable flexibility in this plan and the Gateway provides a safe refuge where the crew could wait out the arrival of a rescue craft to return them to Earth.

Contingencies for safe containment of astronauts are considered essential for sustainable space transportation, something which was lacking with Apollo where there was no hope of rescuing a crew stranded on the surface of the Moon. That will still be a risk with the Artemis programme, although there is in-built flexibility through habitats which could similarly protect crews isolated on the surface awaiting a rescue vehicle from Earth or from a contingency vehicle left at the Gateway for such an eventuality. That is standard practice today at the International Space Station where a second Soyuz spacecraft is always on hand to return a crew to Earth should the assigned Soyuz be found to have malfunctioned.

Blue Origin has emerged as the quiet disruptor, competing indirectly with Virgin Galactic for space tourism, challenging SpaceX for future launch contracts in the medium-to-heavy lift payload categories and now back-up to Elon Musk's Starship for taking astronauts back to the Moon. With the complexity of the SpaceX/Starship plan and its numerous, integrated tanker operations in low Earth orbit to support a single mission, the inherently more elegant architecture of the Blue Moon mission plan and its integrated crew and cargo landers offers an arguably more sustainable approach to long-term lunar surface operations.

Late in the day, Britain bids to join New Space players

he United Kingdom is the only country that has developed an independent capability for building and launching its own satellites and after placing the first one in orbit shut down the programme, choosing to rely on the United States for launching its future satellites into space. The UK also withdrew from its seminal role in developing a European launch vehicle by withdrawing Blue Streak as the first stage for Europa and that story has been told earlier in this publication.

There have been many occasions in which UK space scientists and rocket engineers have been frustrated by a lack of government commitment to the commercial development of space exploration. British aircraft and engine manufacturers have consistently tried to promote independent programmes from which numerous proposals have been put forward to government but the will to provide public funds for these have stalled on the lack of a strong political commitment.

Over time there had been several attempts by British rocket scientists to develop a satellite launch capability, arguably the first credible attempt being MUSTARD (Multi-Unit Space Transport And Recovery Device) conceived and designed during the early 1960s. At the time, everyone in the nascent space industry was talking about the re-use of launch vehicles and to achieve that MUSTARD consisted of three identical delta-shaped vehicles attached to each other and released, or 'staged' away in sequence, leaving the final element to reach orbit. Further cost savings came from only having to tool-up for one configuration, multiple assemblies of which would be available off a production line, thus

avoiding the cost of different stage configurations, each with its own unique rocket motor.

Rocket motors in each element would provide a combined lift-off thrust of 480,000lb (2,135kN) and carry 5,000lb (2,268kg) of payload into space, each element returning through the atmosphere to land like a conventional aircraft. The project was developed during the 1960s by the then British Aircraft Corporation and on paper at least it showed significant reductions in launch costs, with the potential for much reduced prices for customers. At the time there was little hope of commercial development, the costs would have been far too high, and the British government was loathe to invest in technology which it believed had already been developed elsewhere.

In 1970 the government decided not to pursue any further interest and BAC had to stop all work on MUSTARD and enter into discussions with the United States about directing all space-related interests to its post-Apollo programme with the Space Shuttle at its core. The winning contractor North American Rockwell entered discussions with BAC about sub-contract work on building wings for the Shuttle and both governments encouraged that. However, to bolster support for the Shuttle in Congress, NASA obtained agreement from the US Air Force that it would fly its payloads on the Shuttle and perhaps operate a dedicated Shuttle mission of its own. At the time, US laws prohibited any shared involvement with non-US companies for military programmes and even the 'special relationship' could not avoid that.

BELOW • Skylon was another proposed reusable satellite launcher using a radical new form of hybrid propulsion, but which failed in an attempt to raise interest as a national venture. (Adrian Mann)

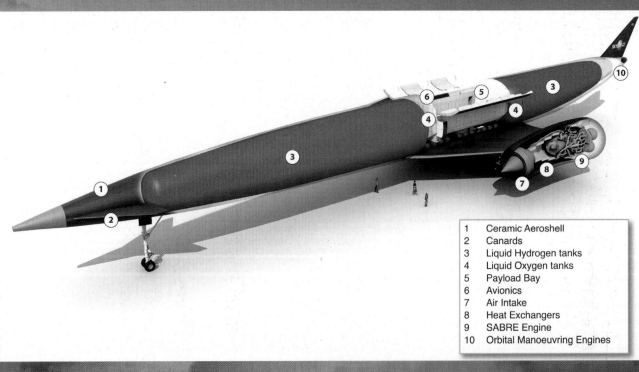

1	Ceramic Aeroshell
2	Canards
3	Liquid Hydrogen tanks
4	Liquid Oxygen tanks
5	Payload Bay
6	Avionics
7	Air Intake
8	Heat Exchangers
9	SABRE Engine
10	Orbital Manoeuvring Engines

It brought a sad end to UK government interest in launch vehicles, and the profits that may have resulted, a time when Blue Streak was cancelled, and Europe left without the all-important first stage for its Europa satellite launcher. But this was not the end of endeavours to get the UK involved. Some proposals have been highly significant, including the HOTOL (Horizontal Take-Off and Landing) concept of the 1980s that proposed a reusable, single-stage concept for placing payloads in orbit.

A successor to HOTOL, Skylon was a refined and more elegantly focused attempt to resurrect the process through a winged aerospace vehicle powered by a hydrogen-fuelled engine using atmospheric oxygen during ascent. Its design lead, Alan Bond, gave it that name from the Skylon structure at the 1951 Festival of Britain exhibition that had inspired him as a boy. The brilliant design ingenuity of Alan Bond, John Scott-Scott, and Richard Varvill converged when the trio set up Reaction Engines Ltd in 1989 after HOTOL was cancelled. Their objective was to develop Skylon and refine the technology into a new and more operationally efficient concept able to take-off and land on its own, versus HOTOL which had a launch-skid arrangement. Propulsion was a development of the HOTOL system and the Sabre engine that emerged had great potential and wide interest both in the UK and the United States.

In some respects, it was a product of its time, the inadequacies of the Shuttle, which was never really an operational load-carrier, each vehicle, and every mission different from the others, bringing urgent demand for more cost-effective means of delivering satellites, spacecraft, and people into orbit. Skylon was a rational solution far ahead of the market it was designed to address. Like the Shuttle, it was not clear that it could be a commercial success due to the limited size of the market at that time. The Shuttle really failed its intended purpose not due to a lack of payloads to launch, but because it was too complex to launch at short intervals. Skylon may have been more suited to frequent re-flights, but we will never know.

All manner of rational analyses would have concluded that Skylon was inappropriate for the time in which it was proposed, as would others examining the dreams of Branson, Musk, and Bezos. Yet their persistence and personal financial commitment created a market that has grown because of what they did to build the traffic models that justified their initial investment – and brought substantial revenue in return. And it is now the governments that were able to see that potential that are benefitting as a result.

It might have been possible for an industrial consortium to have taken Skylon and developed it into a commercial venture, perhaps with government seed money for multiple applications. And that was the intention of REL when it worked up a business plan for 30 aerospace vehicles which it believed it could sell for $1bn each. With these it was believed operators could achieve launch costs of $10m per flight, far below the market fee for conventional satellite and space vehicles at that time.

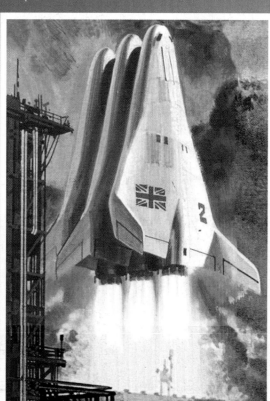

RIGHT • Skylon could have revolutionised space transportation and was unsuccessful largely because the political support was lacking in the UK and Europe. (Adrian Mann)

RIGHT • Orbex is developing the Prime launcher for small satellite applications, this being the second stage with payload fairing. (Orbex)

Reusability was the key and with a payload capability of 33,000lb (14,968kg) to a low Earth orbit the price per flight would have been an order of magnitude lower than anything available elsewhere. Moreover, it would have had the capability of carrying up to 30 people in an appropriately configured and pressurised cargo module. The operational profile of Skylon would have used the air-breathing/hydrogen-fuelled Sabre propulsion system to take off like a conventional aeroplane and reach an altitude of approximately 92,000ft (28,000m) and a speed of Mach 5 where it would switch the dual-mode propulsion to consume internal oxygen to combust with the liquid hydrogen, continue on into orbit and release its payloads.

The complex story of how Skylon was developed and brought to a credible design is long and tortuous, filled with vacillation, a lack of real awareness in the UK government of what they had and how it could have been exploited. In much the way the United States prompted applications of commercial services through initial, fixed-fee incentives helping private companies develop technology which would

bring returns to the federal coffers greater than the outlay invested, so too could Britain by now have had a thriving commercial space industry.

Some seed-money was made available for tests and for development of the Sabre engine but with UK interests wrapped closely in to those of the European Space Agency, everything had to go through European channels. There were several vested interests, not least those in France who had little taste for embracing British projects in rocket propulsion and launch vehicles to challenge their lead in those sectors, over which it had seized control after the UK walked away from Blue Streak leaving Europe to reinvent its autonomous launcher programme.

British Aerospace and Rolls-Royce tried hard to get it approved and eventually found greater interest in the United States, which fed into the broader objective that each company had far greater participation in trans-Atlantic ventures. In some respects, Skylon was a victim of its own inadequacies, in that it was a concept far ahead of its time without the support of either necessary resources or political approval. In the UK, those factors doomed the project before it was able to gather partnerships that could have carried it along to at least the prototype stage.

When presented as an alternative to next-generation expendable Ariane launchers, both HOTOL and Skylon suffered the same fate, and which prevented the UK from developing its own, independent launch system. But the skills and ingenuity of British scientists and engineers continued to grow on projects linking UK capabilities with the rise of the European Space Agency (ESA) in the mid-1970s where received work is proportional to the investment paid in to the general pot by national governments.

Orbex Rising

Today, 22 countries are members of ESA, with France the dominant financial contributor at 24.5%, followed by Germany with 21.2% and Italy with 14.1%. The UK contributes a little under 9%. Membership is not solely for EU countries and the 50-year history of ESA places its formation long before the present structure of the European Union. In Britain, the UK Space Agency (UKSA) is sponsored by the Department for Science, Innovation and Technology with its primary objective being to 'boost UK prosperity'. As such, it promotes British space industry and represented companies and organisations nationally and abroad, with a small pot of money to stimulate projects. One of those campaigns has been to support spaceports developed by private organisations.

Virgin Orbit (see page 54) took up that opportunity with an agreement through local councils to send Launcher One

BELOW • The Orbex headquarters at Forres, Scotland from where the small-satellite launcher business is managed. (Orbex)

rockets into orbit from Newquay, but the company halted operations after a failure in the one attempted flight. Others are being more successful by developing sites for vertically-launched rockets carrying small satellites. One such is Orbex, founded as Moonspike in 2015 and renamed Orbital Express Launch the following year with the objective of launching Prime rockets carrying satellites weighing up to 330lb (150kg) to near-polar Sun-synchronous orbits from a site on the coast of northern Scotland.

With headquarters at Forres, Moray, Orbex will fly Prime, a two-stage rocket powered by propane and liquid oxygen, so-called 'green' propellants, from the Sutherland Spaceport on the A' Mhòine peninsula. The design of the rocket anticipates recovery of the first stage using parachutes and deployable petals to slow it down as it descends. The design is simple and efficient, with the outer casing fabricated from carbon-fibre materials. The original intention was to share the spaceport with Lockheed Martin with interests in the Electron rocket operated by Rocketlab in New Zealand (see page 118). That rocket has a different propellant combination and would require a second launch pad, while a licence for only one has been granted by the Scottish government.

Orbex has investment from several companies wanting to share in the burgeoning market for small satellite launches, supplemented by a small fund from the Scottish National Investment Bank. Market conditions for small satellite launchers are right, now that Arianespace has retired its Ariane 5 and is facing delays and uncertainties with its successor, Ariane 6. Orbex wants to get its foot in the door before upscaling plans but expects the

opportunities to arise for a slightly larger development of Prime to widen its appeal to a broader market.

Lockheed Martin's interest in launching Pathfinder rockets from the UK has boosted activity at the SaxaVord spaceport on the Lamba Ness peninsula, Unst, Shetland Islands. British company Skyrora successfully signed a deal for the use of this location for launching satellites into space and that is the place where Pathfinder could fly from for customers across Europe. With flight paths directly out across the sea, it is geographically ideal for sending polar-orbiting satellites into space.

Some objections were lodged because work to prepare the site would destroy a Chain Home radar station from World War Two, which is now a national monument, but opposition was withdrawn as it was said to be in the interests of the local economy. With agreement approved, the German firm Rocket Factory Augsburg (RFA) has obtained agreement to launch its own rockets from SaxaVord and late in 2023, the UK Civil Aviation Authority granted licences for up to 30 launches a year from this location. With headquarters in Augsburg, Germany, the RFA One rocket will be capable of lifting 2,866lb (1,300kg) satellites into orbit from SaxaVord as well as from other locations.

While investment in big rockets for heavy-lift satellites and spacecraft is now firmly in the hands of commercial space operators, primarily in the United States, the small satellite market is growing fast and in need of rockets to get its products into orbit, a surge fuelled by increased use of space supporting everyday needs for ordinary citizens.

ROCKET LAB

Up high from down under with rocket flights

In what many may regard as one of the more unusual journeys into a career involving rockets and space engineering, New Zealander Peter Beck trod a varied path to management of Rocket Lab, one of the most successful start-ups in commercial space activity in the last 20 years and one which is moving quickly to enter the medium-launcher market as well.

Brought up in Invercargill, New Zealand, Beck was drawn to mechanical devices and spent much of his youth working on cars, adding a turbocharger to a Mini and playing around with water

rockets. It was only when he began work as an apprentice tool-and-die maker that he harnessed natural skills and directed his talents into design and crafting of rocket packs for bicycles, scooters, and jet-packs before working as an engineer on a yacht project in New Plymouth on the North Island.

Beck liked to take things apart and improve them, at one point buying a redundant Williams turbofan engine retrieved from a decommissioned cruise missile. He worked for an industrial research company from 2001 developing smart materials and superconductors. It was there that he met the business entrepreneur Stephen Tindall who would be an investor in his Rocket Lab company, founded in June 2006 after Beck visited the United States where he became convinced there was a market for small satellite launchers. Appropriately named, Mark Rocket met Beck and together they attracted other investors which included Tindall, Vinod Khosla, and the New Zealand government, which granted seed money.

Beck was actively involved in experiments with sounding rockets firing instruments high into the atmosphere and recovered by parachutes. On November 30, 2009, his 19.7ft (6m) tall Atea-1 sounding rocket weighing 132lb (60kg) became the first private rocket from the Southern Hemisphere to reach space. It achieved an altitude of 100km following launch from Great Mercury Island off the Coromandel Peninsula, a private island owned by the banker Michael Fay, another Rocket Lab investor. The rocket fired for 22 seconds, reaching a speed of

RIGHT • One of two NASA TROPICS satellites goes into orbit from the New Zealand launch site in 2023. (Rocket Lab)

BELOW • Rocket Lab's Peter Beck poses with NASA deputy administrator Dr Dava Newman during a meeting to discuss future cooperation. (NASA)

3,100mph (5,000kph) before separating the 4.4lb (2kg) instrumented top section which was never recovered. The first stage was found by a local fisherman.

Much of Rocket Lab's success began when it started to develop its interests in the United States. In December 2010 it was awarded a contract from the Operationally Responsive Space Office (ORSO), formed three years earlier at Kirtland Air Force Base, New Mexico, for a study into low-cost launchers for putting CubeSats into orbit. These are very small satellites, each a cube 3.9in (10m) wide on each side and weighing no more than 4.4lb (2kg). They are cheap to produce and can support any function for which they are programmed, accessible for use by fan-clubs, schools and organisations looking to ride piggy-back on conventional launchers or in clusters on low-cost rockets. The ORSO wanted to stimulate the market and encourage greater use of CubeSats.

The ORSO agreement provided access to NASA and availability of its facilities and resources for development, as well as gaining Rocket Lab valuable experience with how the agency operates and how to work with it. By 2013, the company had re-registered in the United States, converting its original company in New Zealand as a subsidiary of its American prime. By 2020 it had moved to facilities in the Long Beach, California, area where it now manufactures its rockets and carries out research and development.

An early objective for Rocket Lab was the design and operation of a small satellite launcher and Beck was soon organising such a programme with his Electron rocket, the first flight of which took place on May 25, 2017, from a launch site at Mahia, New Zealand. It reached a height of about 139 miles (224km) when telemetry was lost, and it was destroyed by the range safety officer. The second flight on January 21, 2018, carried CubeSats into orbit and was followed by two more successful launches that year. Six flights were completed in 2019, all successful, including the first for the US Air Force, and one failure out of seven launches was logged for 2020. Again, there was a single failure in six attempts in 2021 and one failure in nine launches during 2023.

The first flight from the US launch site at the Mid-Atlantic Regional Spaceport (MARS) in Virginia took place on January 24, 2023, when the 33rd flight of Electron launched three satellites for HawkEye 360, a geospatial company based in Herndon, Virginia. The second launch from this facility followed with the next flight on March 16, 2023, with the launch of two satellites for Capella Space, a company which takes declassified synthetic aperture radar data and adapts it for commercial use. The third flight from MARS was on March 21, 2024, which was also the first Rocket Lab launch for the US National Reconnaissance Office (NRO) from the United States. Four previous flights for the NRO had been from the New Zealand launch site. If all flights take place, Rocket Lab could launch 25 Electron missions in 2024 and it has a full order book for 2025.

ABOVE • Electron launches a satellite for the US National Reconnaissance Office on January 31, 2020, from its New Zealand pad. (Rocket Lab)

Moving its business base from New Zealand to the United States opened possibilities it would not have had under its original registration, US government constraints on sending certain work outside the USA, be that products or services, otherwise prohibiting a lot of the launch contracts Rocket Lab has received by being in California. This is a consideration for any commercial launch provider seeking to exploit the surge in space activity. It is a market worth pursuing. The global space economy has a $470bn turnover growing at almost 10% per annum, with revenue of $279bn. The amount of tax paid by the global space industry is three times the amount of money invested by taxpayers. A considerable portion of that business is in the launcher industry with a growing segment from the commercial operators and providers.

Electron Charge

Rocket Lab's Electron is a partially reusable, two-stage launcher using kerosene and liquid oxygen for the Rutherford rocket engine of which nine power the first stage generating a lift-off thrust of 43,000lb (191.2kN). A single Rutherford powers the second stage delivering a vacuum thrust of 5,800lb (25.8kN). Electron also allows for a Photon kick stage to insert payloads directly into orbit with a thrust of 27lb (120N) or in more recent variants a modified stage with a thrust of 90lb (400N).

Under a normal flight profile, the first stage fires for two minutes 35sec before separation at an altitude of 48 miles (78km) after which it falls back to Earth and currently work is underway to recover it with the aid of an inflated parawing. Ignition of the second stage occurs seven seconds after separation and, at a height of 51 miles (82km), the payload fairing is jettisoned 22 seconds later at an altitude

of 78 miles (126km). Engine cut-off occurs after a burn duration of six minutes 13sec and separates from the kick stage and its payload four seconds later. Flight loads during ascent are relatively benign for an uncrewed vehicle, limited to 7.5g in compression and plus or minus 2g for lateral loads, mostly during stage separation.

The technology in Electron is both advanced and sophisticated, the Rutherford motor being, uniquely, pump-fed via brushless DC electric motors and high-performance lithium polymer batteries with a 90% efficiency. This is also the first engine to use additive manufacturing in all its primary components including the regeneratively-cooled thrust chamber, injector pumps and the main propellant valves. Engines for both stages are identical, differing in having a larger expansion nozzle for the second stage for optimum efficiency in the vacuum of space.

The Electron has a carbon composite structure which reduces weight by as much as 40% over standard aluminium and the optional use of launch facilities on different continents has greatly increased the market potential. From Launch Complex 1 on New Zealand's Mahia Peninsula, Electron can reach orbital inclinations between 49° and 120° while trajectories from Launch Complex 2 at the Mid-Atlantic Regional Spaceport in Virginia access orbital inclinations from 8° to 60°. Not that the vehicle cannot access other inclinations required by customers, but those would involve a turn in the ascending trajectory to align with the required flight path which would require additional propellant and reduce the payload capability.

Rocket Lab has made great strides to design and integrate its own avionics and flight computer systems, all assembled and tested in-house. Computing nodes use field programmable gate array architecture which preserves commonality for all the basic functions while allowing bespoke customisation for individual payloads and performance requirements. This is akin to the open-architecture avionics suites now designed in to modern combat aircraft, a practice which really began in the aviation industry with the Lockheed Martin F-22 Raptor combat aircraft. As now required by the FAA in the United States, a certified flight termination system allows automatic or ground-commanded destruction in the event of a failed rocket running amok and threatening populated areas.

Rocket Lab have developed a highly efficient design, development and production model which benefits from vertical integration, where all elements of the rocket and its propulsion system are manufactured in-house rather than from a range of sub-contractors and ancillary suppliers. It is difficult to over-emphasise the benefit this approach has had on efficiency, cost-reduction, and speed and, as with SpaceX and Blue Origin, in-house 'ownership' of all the engineering and technology, design, manufacture, modification, fabrication, and flight support activity has delivered progress at pace.

The vertical integration used by these entrepreneurial organisations is impossible to incorporate in the way existing, major aerospace manufacturers operate today. The distribution of subcontracts across the 50 states is a formal part of any US government programme where taxpayer money is involved. In its briefing before Congressional committees, NASA has to show diversity of work across as many political regions as possible, so as to attract support from the electorate for representatives and senators. This encourages the major aerospace companies to spread their work out across many different facilities, subcontractors, and suppliers, increasing costs, extending development cycles, and contributing to delays.

Rocket Lab is a classic example of how quickly smart organisations work more efficiently than those working faster, respond more quickly to design changes, incorporate modifications and performance-enhancing techniques more quickly and gain added business as a result. Since its first launch, almost each successive flight has incorporated some improvement in assembly, systems, or operating modes. This is a pattern displayed by other companies and organisations using vertical integration and is evident in the responsiveness

and performance described for commercial space operators throughout this publication. It is admirably demonstrated by the enhanced performance of Electron achieved through rapid adaptability, by any objective judgement an exemplar for New Space and its commercial activities.

The initial payload capability of up to 496lb (225kg) placed in a Sun-synchronous orbit has now grown to 661lb (300kg) with a greater range of optional services in the electrical and electronic support for payloads, be they individual satellites or a series of smaller instruments. In 2023, Rocket Lab also offered a sub-orbital version of Electron which is capable of delivering 1,543lb (700kg) to a ballistic trajectory for several minutes of weightlessness, the first such launch having taken place from Launch Complex 2 on June 18.

The future for Electron is healthy and the possibilities for other small satellite operators promises added business through the general expansion of this sector. Advances in electronics and the miniaturisation of components and systems makes possible the same services that previously would have required large and costly spacecraft. While there will always be a demand for big satellites placed in high orbits, there is an increasing value from smaller payloads in lower paths and for that Rocket Lab is a clear example of real opportunities for a new generation of space flight operators.

ABOVE • Here unfurled, an advanced composite solar sail being prepared for a launch aboard Electron to test the ability to use the pressure of the solar wind in space to move space vehicles. (NASA)

LEFT • The Spire satellite facility in Glasgow, Scotland, preparing satellites for launch on Electron. (Quentin Gollier)

PRIVATE
MOON LANDERS

Global players are racing to compete for lunar resource materials

With strong commercial development of satellite services, launch vehicles, even space tourism, companies and entrepreneurs are looking to the Moon as the next step. The key to expanding commercial operations is the Artemis Accords, a series of bilateral agreements signed on October 13, 2020, and now sealed with 40 countries. It sets out an international charter for the exploration of the Moon under the NASA Artemis programme, which expects to place the next humans on the lunar surface sometime within the next three or four years.

The Accord establishes a framework within which international governments, organisations and companies can participate in the full commercial development of lunar resources and operate under a code of law which conforms to the 1967 Outer Space Treaty prohibiting a country or organisation from laying claim to any part of the Moon. The United Sates sees itself in a race with China to be first on the lunar surface this century and the international Artemis Accords are one way of inhibiting unilateral or national action in claiming rights over minerals and resources believed to be present on the Moon.

But governments cannot afford to develop a lunar-based economy. It needs the commercial world and private enterprise to fund the next steps and the financial return many believe to be in harvesting lunar resources is expected to pay for that. To help industry get there, in November 2018 NASA launched an inducement by starting the Commercial Lunar Payload Services (CLPS) programme. Nine companies were invited to bid on contracts for robot landers and roving vehicles to survey areas where Artemis Moon landings are planned to take place near the lunar South Pole. It is there that the greatest interest lies in finding water ice and other potentially interesting resources.

CLPS contracts went to Astrobotic and Intuitive Machines. Founded in 2007, the first Astrobotic launch occurred on January 8, 2024, with the flight of Peregrine aboard the inaugural flight of the Vulcan rocket developed by United Launch Alliance to replace its Delta and Atlas launch vehicles. Peregrine weighed 2,829lb (1,283kg) and was assisted in its design by Airbus Defence and Space, the European manufacturer famed for its aircraft and spacecraft. The flight began with separation from the Centaur upper stage at 50 minutes 26sec after

lift-off to start a long, looping trajectory that would have had it arrive at the Moon 46 days later on February 23 after a series of manoeuvres using on-board thrusters.

About seven hours into the flight a problem arose with the stability of the spacecraft, controllers finding it hard to keep its solar panels pointing at the Sun. Power began to drain and it was clear that there was a leak of propellant slowly spinning it around. It was only a matter of time before the propellant essential for landing on the Moon ran out and that sealed the fate of Peregrine, which was still in a looping orbit of the Earth. Reconciled to its fate, controllers used the last remaining propellant to change its path to one which brought it back down through the Earth's atmosphere on January 18, far short of its objective.

The Intuitive Machines IM-1, a Nova-C class spacecraft named Odysseus, was next up, launched by Falcon 9 on February 15, 2024. It became the first US spacecraft to soft land on the Moon in more than 51 years when it touched down, albeit resting at an angle of 30° and tilting its antenna away from Earth making communications difficult. That compromised its performance, but it was the first commercial lander to reach the lunar surface. The second mission, IM-2 is expected to launch in the fourth quarter of 2024 carrying a drill to search for ice and an independently operating 'hopper' to move around and explore craters. It will be followed by IM-3 in early 2025 carrying a small roving vehicle to investigate local features.

Intuitive Machines is planning further lunar landers including one capable of carrying 1,100lb (500kg) of scientific equipment

TOP • Seen in false-colour rendition, the lunar hemisphere displays the wide range of surface materials to attract industrialists of the Lunar Age. (NASA)

to the surface and a more advanced version with 11,025lb (5,000kg) of equipment. This scaling-up will be important for complementing the Artemis flights carrying astronauts to the surface and pre-positioning equipment or delivering specific tools and instruments to sites already established.

Astrobotic is hoping for better luck than it had with Peregrine when it launches its Griffin lander, a larger version carrying a Viper roving vehicle to investigate shadowed regions of the South Pole looking for deposits of water ice. The failure of Peregrine prompted further examination of the engineering design and with external analysis there have been some changes to the spacecraft and rover. With help from NASA, Astrobotic is developing a new class of lunar rovers which will have a standardised architecture capable of receiving a wide range of different instruments and electronic controls for different

applications. The idea is to take the CubeSat concept and apply it to multifunctional robots, available off-the-shelf.

Commercial lunar exploration projects take innovative ideas and inject competitiveness to spur designs which achieve unique capabilities. They invariably carry NASA science experiments but attract different concepts and these could lead to big returns for commercial companies. A lot of this work is carried out by universities and research institutions specialising in particular areas which connect with broader requirements. It is inevitable that some will fall by the wayside, but the stimulation for creativity is high.

Many of the lunar materials would support sustainable activities at the surface, including oxygen, water, hydrogen, and metals. Many areas on the Moon still preserved from the earliest time of its formation, places known as the Highlands, contain anorthite which can be used as an aluminium ore from which can be obtained calcium metals, oxygen, and silica. Materials can be separated on site with the use of potassium fluoride to part out the raw materials. There are many ways in which oxygen could be extracted from the lunar dust by heating it to high temperatures and introducing hydrogen gas. The Clementine lunar orbiting spacecraft discovered large quantities of hydrogen in soils near the lunar poles.

Other possibilities exist, including the use of various processing techniques for sintering, hot-press and liquefaction techniques with production of glass and glass-fibre from surface materials. In research experiments, basalt fibres have been produced from simulated lunar surface materials. Despite the name, rare-earth elements are relatively abundant on Earth where they can be hard to find and apply due to political constraints. They can be found on the Moon where access is not dependent on such limitations and where mining them could bring rewards to more than compensate for their extraction. They are vital for optical, electrical, magnetic, and catalytic uses on Earth and are considered one of the rich prizes for potential investors.

Another lunar resource is helium-3, which could be useful in nuclear fusion reactors, but they are still in the experimental stage and offer little commercial advantage in the short term. Not so the availability of minerals from the extensive basaltic lava flows on the surface, the darker patches as viewed from Earth. These contain elements and minerals that can be broken down to produce titanium and in 2020 the European Space Agency contracted Metalysis, a company specialising in solid state metals and powders, to develop a unique process for processing lunar materials to produce that with oxygen as a secondary product.

Mining the Moon

In May 2024, one of the most influential think-tank strategy organisations in the US, the Defense Advanced Projects Research Agency (DARPA) issued a 10-year plan for exploitation of the Moon's natural resources. Not by big government, Apollo-style, programmes paid for by US taxpayers, but rather by providing a framework for commercial companies, private organisations, and wealthy entrepreneurs to follow the lead set by launch operators and New Space investors and set course for the Moon to mine its resources. Titled LunA-10, (10-Year Lunar Architecture Capability Study), the report's conclusions and its associated conference presentations bring several years of intensive work to bear on the size of the commercial market and the number of people involved.

Mining the Moon is believed to be the major driver toward future commercialisation of the space programme, transferring government programmes to the private sector and increasing the number of companies involved. In looking to the future, it sizes the scale of the potential market and presents it as the next great leap in business opportunities, with government getting out of the way. Legislation, regulatory controls on safety, and international agreements would be a strong part of the new race for celestial resources to replace or consolidate depleting stocks of rare-earths and vital minerals for a future envisaged by politicians but impossible to achieve with existing supplies or Earth-based materials.

The degree of interest measured by commitments already made on lunar basing, construction, mining, and transportation costs leads to some startling conclusions. Companies and private organisations across the democratic world have already begun planning to invest in the new space race, one controlled by the opportunities for industrial activity supporting existing requirements for these materials. Governments can no longer afford to finance big space

ventures. As noted previously in this publication, projects which at one time during the Cold War would have been seen as essential pillars of national prestige and as useful propaganda to show political or ideological superiority, are now replaced with private sector participation to pay for elements unaffordable to governments.

The NASA Artemis Moon programme is a classic example of how many nations around the world are providing spacecraft, rockets, modules, and people for what is still badged as a US venture. Instead, the United States is now only the organisational lead and the political instrument for harnessing the resources of both public and private companies to make it happen. Increasingly, that will be the model for future development of what is now considered to be as important a step in world economic progress as the great trading companies of a former age which funded unprecedented leaps in science and technology.

To support this new industrial revolution, the number of people involved in these projects is expected to increase greatly over the next 20 years. The DARPA study takes the Artemis programme and considers the current commercial activity - either in development or operating today - and estimates that by 2030 there will be 40 people working on the lunar surface, either intermittently or continuously. That number is expected to grow to 200 by 2035 and reach 1,000 by 2040. But this is only on existing plans and projects from work already in development. It could be a lot higher if the commercial operators such as SpaceX and Blue Origin expand their capabilities – and there is no sign that they will not.

There has never been a time when so many industries are showing keen interest in applying their sectors to the possibilities opened by routine and reliable space transportation. Projections are based on precedent, the example of SpaceX's penetration of the launcher market being particularly appropriate. Twenty years ago, it was a billionaire's plaything. Now it launches the majority of the world's satellite and space traffic into orbit with an order book larger than any previous operator and is the cornerstone of NASA's plan to get

ABOVE • Odysseus was the first lander built by Intuitive Machines and became the first commercially developed Moon lander to reach the surface. (Intuitive Machines)

LEFT • Astrobotic's CubeRover during development and test, prelude to a new generation of mini-rovers. (Astrobotic)

ABOVE • *NASA tests an automated lunar resources extraction machine under simulated Moon surface conditions. (NASA)*

back on the lunar surface. Now, other industries see ways in which their niche own skills are relevant.

Construction companies are examining the way building infrastructure and site facilities for habitation, energy production, experiment facilities and mining activities can be provided on the lunar surface. Even architects are commissioning studies from specialist engineers on what is required to construct a sustainable habitat on the Moon. Food production on the Moon will be an important way of minimising the need to bring stores from Earth, with agricultural companies now examining how plants grow in lunar dust, an activity that has been underway experimentally for many years with positive results. Pharmaceutical industries are also looking at medicinal requirements for Moon workers and synthesising products which combine curative effects to minimise the risk of illness.

More directly associated with the physical activity of mining the Moon, companies carrying out this work on Earth are looking at the requirements for machines, processes and the separation of materials extracted from the terrain. Transportation across the surface will require new vehicles to carry people, tools, and materials around and manufacturers are competing already for contracts to design and produce wheeled vehicles for operation on the lunar surface. Adverts are beginning to appear from car companies proclaiming their stake in the future, moving workers around on the Moon, and NASA is examining bids for development work, seeded by government but taken over by commercial manufacturers.

Energy production on the lunar surface is essential to providing the power for sustaining colonies of workers, scientists, geologists, and engineers. The development of mini nuclear power stations may point the way toward how such equipment could be installed on the lunar surface. NASA is already looking to see how nuclear power can be the power-base for sites on the Moon where at best they only get two weeks of sunlight each month. And there are several places in almost continual darkness where scientists believe useful materials may be found. The use of nuclear power to power the lunar worksite and habitats is more than an option, it is a necessity.

Much of the physical work extracting materials and natural resources could be undertaken by robots, coring out and separating lunar materials for compression and packaging for dispatch to Earth and specialised suppliers are looking at the way existing equipment could be adapted for work on the Moon. There are no unique challenges in this sector and industrial mining companies are currently working with government space agencies to gauge the level of workload required.

Communications too would be essential for industrial activities, but the absence of an interconnected system would attract telecommunications companies on Earth, adapting their equipment and systems for use on the lunar surface and with spacecraft in orbit, places such as the Gateway which will be the terminus for Earth-Moon access. An early requirement would be a constellation of satellites to act as relay nodes much like cell-phone towers on Earth connect smartphone users. Operations conducted on the far side of the Moon which never faces Earth would only be possible through a dedicated lunar communications system perhaps consisting of telecommunications satellites in highly elliptical paths connecting operations at the surface with bases on the near side of the Moon and with mission control on Earth.

Late in 2023, NASA began a search for proposals on how to develop and standardise a lunar clock in which operations around and on the surface of the Moon would have its own time, separate from the Universal Time Clock (UTC) aligned with the old Greenwich Mean Time standard used throughout the world for transport and clock reference. This would take account of the greatly extended 'day' which to workers on the Moon would be one Earth month from one sunrise to the

RIGHT • *Intuitive Machines submitted a lunar vehicle design in cooperation with several companies associated with the car and aviation industries, including Boeing which built the first Lunar Roving Vehicle used on the Moon by Apollo astronauts in the 1970s. (Intuitive Machines)*

next. Routine working schedules would have to accept that the human body has a biological clock, a circadian rhythm which is vital for health. In that regard, some synchronisation with Earth time would be necessary while retaining a sense of relevance to the local day/night cycle on the Moon.

Building it Big

In assessing the amount of material sent from the Earth to the Moon to sustain a working resource base, engineers work to an average 45 tonnes of mass required per person each year. By 2040, if projections are correct about the number of people on the Moon that would require an upload of 45,000 tonnes per year. To place this in context, it would call for 500 annual SpaceX Starship flights – which is well within Elon Musk's expectation of a maximum capacity for planned infrastructure of three launches per day.

But this is a high number and would add to the recurring costs of supporting a lunar-based economy. Far better to expand the in-situ surface infrastructure and manufacture as much of the consumables for supporting human activity as possible on the Moon. It is highly probable that a developed lunar surface facility could support itself to the equivalent of 35 tonnes per person per annum, relieving the Earth-Moon transport system of that amount. Nevertheless, delivering mining and support equipment and returning mined resources to Earth would support a lunar transportation value of $751bn between 2036 and 2040, an average of $150bn per year over that period.

Today the global launcher market is worth $14bn. With a more than tenfold increase in turnover, launch prices would topple and competition to enter the rocket market would grow, profits soaring and new entrants increasing. Investment is likely to grow quickly and already in 2024 there are signs that this is happening. In the projected period 2020 to 2040, the total value of lunar transportation over that 20 year period would reach $1,250bn, of which it is anticipated that commercial companies in the United States would get 51% of the total share with China and Japan getting 31% and Europe along with the rest of the world carving out $18%.

The EU has less of an incentive for private and commercial operators. Like the UK it has a top-down approach toward business opportunities and higher financial charges for private space operations. In the United States there is a move to get government out of the way and have it focus on regulatory standards and that is unlikely to change over the assessed

period. The European Space Agency will have a significant role to play in the provision of lunar resource development and it is already a major player in NASA's Artemis programme but work on that is contracted to major European aerospace companies and paid for by governments at constrained profit margins leaving little incentive for privately-owned contenders.

The commercial development of space has reached a tipping point where profits already flow from initiatives exploited outside government programmes and grown in private companies around the world. Politically polarised countries are on a different path, China already invests large amounts of government money in seeking to establish a pre-eminent position in the new space race with Russia now playing a much less significant role. China professes to support entrepreneurial space projects privately owned and funded by the country's banks, but all its national companies and organisations exist at the behest of the state and not on the basis of their financial viability.

In the next 20 years, New Space will have replaced much of the established order characterised by managing space exploration as a political tool for prestige or power. The commercial development of technologically sophisticated consumer goods and the increased popularity of electronic items and related devices, of TV and data-driven services and of automated navigation, has dramatically expanded the market and driven up the demand for satellites. That alone has created the wealth that drives entrepreneurs to make the satellites, space vehicles, and rockets to launch them, pushing humanity toward a multi-planet future.

ABOVE • A work site at the Moon's South Pole with surface materials covering habitats and work facilities for protection from radiation and micrometeoroid impacts. (NASA)

LEFT • A lunar surface base would use robotic and crewed vehicles to establish an industrial site. (NASA)

FUTURE
FLIGHT

From the Moon and Mars to worlds beyond

It is very likely that commercial space will be the weather-vane on where to go after the Moon and Mars, although the challenges of colonising planets further away are beyond anything that can be faced today. The rest of the Solar System beckons, but the conditions far away from the Sun are cold, formidable, and inhospitable. None of the outer planets have defined surface features due to their atmospheric envelopes that only gradually become solidified at lower altitude. But they do have moons, many of which draw attention due to their bizarre and curiously unusual features and they may be visited by a passing Starship or some other such mass people-mover cruising beyond Mars.

More likely perhaps is the possibility that future activity may refocus attention much closer to Earth, orbiting colonies, vast artificial cities floating in space far above the planet, or giant solar array farms beaming to Earth on microwave links electrical energy converted from sunlight. These Solar Power Satellites have been proposed many times and could offer levels of power unattainable from the surface of the Earth, either from nuclear power sources or from sustainable means such as wind turbines, tidal turbines, or solar cells.

The principle is basic, operating from extensive arrays of photo-voltaic cells laid out across giant structures, in sections uplifted from Earth. Assembled in space and laid out across several hundred square kilometres, several millions of solar cells would be continuously exposed to sunlight. Converted to microwave frequencies for transmission down to the surface of the Earth, the electrical energy would be received at giant rectennas and converted back into usable power for distribution along existing power lines. Such a system of several giant power satellites strung out along the equator at geostationary altitude could provide many times the energy required for a future where at present demand exceeds supply.

Such platforms could eliminate the need for any power-production system on Earth and provide virtually limitless amounts of energy. Today, there is insufficient capacity to accommodate all the uses that electrical power is required for. For a future in which demand will only grow, such giant engineering projects appear reasonable in comparison. So it may be that future space commercialisation will be run in part by electricity supply companies assembling, servicing, and maintaining an off-planet grid, beaming down energy to giant collectors on the ground, at microwave transmission levels less than those employed today for cell-phones and wireless internet services.

It is difficult to engage with the immensity of such a colossal project but two hundred years ago it would have been impossible to imagine the world today in which up to 20,000 aircraft are in the air at any one time, carrying almost

three million people through the skies. A world in which power is provided from orbiting power stations in space may not be so far-fetched. But it would require massive amounts of material ferried into space and in the past that has been impossible to achieve. Now, with launch rates for conventional satellites and spacecraft soaring and rockets on a scale never built before, it becomes a very real possibility.

Floating Cities

While there may be no second Earth beckoning in our Solar System, the possibility of establishing a space colony assembled over time with material mined from the Moon and the asteroids has been proposed since the early years of the Space Age. It is an idea that defies logic. Why would people want to do that? It was first brought to formal consideration by Gerard O'Neill in the 1970s when all things seemed possible. Men had landed on the Moon; NASA was building the Space Shuttle and fiction was blending with fact.

O'Neill recognised that the United States had in hand all the tools it needed to establish a space manufacturing industry, using materials from the Solar System to create an independent, fully self-sustaining space habitat large enough to support several hundreds of thousands of people. In his seminal book *The High Frontier: Human Colonies in Space*, O'Neill seeded an idea that has never gone away.

RIGHT • Space colonies as imagined by Gerard O'Neill, rotating cylinders providing artificial gravity for a space-based city populated perhaps by several million people. (NASA)

FAR RIGHT • Trips to Mars may require artificial gravity with electrical propulsion and power from solar cells or nuclear-electric generators. (NASA)

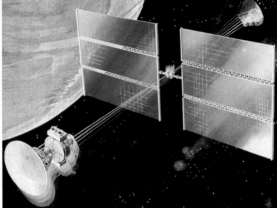

Envisaging a colony based on rotating cylinders providing artificial gravity, he proposed a 'second Earth' which appealed to many people in different ways.

This was a time when much concern was being raised about the limitations of the Earth's environment and the profligate use of its finite resources. President Jimmy Carter was a strong supporter of environmental studies and encouraged people to examine the way technology and space-based capabilities could contribute toward a more efficient and sustainable use of the planet's resources. Followers of the 'green' revolution saw in O'Neill's idea one way of alleviating the surge in global population, seeing a second Earth as a plausible way to plan a future in which urban density could be constrained by providing an optional living space off-planet.

As remarkable as it may seem today, O'Neill's ideas took hold and technical analysis of how floating cities in space could be assembled was given attention in universities and at NASA. Anticipating the amount of material required, some engineers proposed giant launch vehicles which even in the unrealistic optimism of the 1970s attracted design studies to make it possible to carry out such work. O'Neill's space colonies required the same amount of material as the giant solar power satellites and there was a common thread running through these studies – the need for bigger rockets, cheaper access to space and a genuine space-based economy. Which is where the commercial space advocates came in.

Bigger, Better, Further

Today, startling concepts such as O'Neill's floating cities are once again the talking point of space advocacy groups around the world. Such ideas are not for the faint-hearted and certainly not for governments to fund.

BELOW • Travelling the space lanes of the Solar System, future tourists get a stunning view of Saturn's rings. (SpaceX)

But existing companies now offering New Space services in a self-sustaining, self-funding economy are mindful of what space colonies could provide. Concerns about environmental disasters, global conflict or the attentions of an incoming asteroid seeing a solution in that concept. It is not only getting the attention of futurists but also the big players in today's commercial space world. Jeff Bezos has begun to formally discuss O'Neill colonies and that is getting attention.

For the more immediate future, commercial space operations have yet to extend tourism to the Moon and the planets, although SpaceX is already sending private citizens to the International Space Station and preparing for sightseeing trips around the Moon. It can only be a matter of time before those offers will open tour routes to Mars and perhaps to wonders beyond. How magnificent it would be to fly alongside the rings of Saturn, swoop low over erupting geysers on Jupiter's moons or glide serenely across the giant ice worlds of the outer solar system. Such trips are not possible today, but the dream is alive that these may be achievable within the next 20 years or so.

In the more immediate future, new possibilities are opening elsewhere and the New Space of commercial activity that underpins so much that humanity relies on today is growing at unprecedented rates and to new levels. It would not be possible to achieve many of the projects now managed by leading space agencies around the world were it not for the contribution made by entrepreneurs and innovators. Free from government constraints on planning and funding, they exploit technical opportunities for a self-sustaining industry supporting space projects for human benefit as well as the expansion of exploration far out in the solar system.

Nobody could have foreseen the transformation in space applications brought about through private enterprise and the vision of innovators, engineers, and investors. What is now possible through commercial space operations goes hand-in-hand with many national objectives that underpin modern society, international partnerships and the enduring ideal that space should be free from conflict, territorial claims, and over-exploitation of its abundant resources.

We have much to learn from the mistakes of the past on using the resources of this Earth, but the efficient and benevolent use of space technology can bring peaceful benefits to the lives of Earthlings unimagined only a few years ago. It can also inspire and elevate the spirit as humans continue to explore the natural wonders of nearby worlds, stimulating new generations to join the journey and pave a path to the stars.